AF874493

Originally published in 1991 by Allan T Condie Publications.

This edition published in 2006 by Japonica Press
Low Green Farm, Hutton, Driffield YO25 9PX.

ISBN 978 1 904686 10 1

A catalogue record for this book is available from the British Library.

Published by Japonica Press
Low Green Farm, Hutton, Driffield
East Yorkshire, YO25 9PX
United Kingdom

Original design layout and editing by Allan T Condie.

INTRODUCTION

In my pre-school years there was no other tractor worth considering than a Fordson Major. I did not know its proper title 'E27N' until many years later. Indeed, to most the "Fordson Major" was its name. To the dealers it was the 'E27N', and perhaps, through the influence of my writings it is by those code initials that the model is best known in the 'nineties'.

If you want to catch a cold, now is the time. The E27N is not exactly the user friendliest of models, with you perched up in the air exposed to all the elements. Nevertheless the variety of this model can be gleaned from a dip in this publication. P6(TA) diesel engined models are much saught after, but beware - Perkins engines are expensive to repair!

One of the handicaps of all post 1936 Fordsons, including Majors, is the fact that inevitably there were more with the 4.3 transmission around. The red or green spot does not signify the top gear of the tractor, but the 1st/2nd ratios. But again a look in this publication will reveal all, as not only can you check this data easily, but also the exact date of many of the components on the E27N can be ascertained from the casting codes.

What this 'Special' has tried to do is to bring the reader, who is most likely to be someone who is going to set about restoring an E27N, as much information as possible about the changes to the model from 1945-52. Your author is fortunate in having an almost complete set of E27N service and parts literature, and it is from this, along with nearly thirty five years of observation of the species, that brings you this, the most complete E27N treatise yet, onto your bookshelf.

Carlton, January 1991. **Allan T. Condie.**

ACKNOWLEDGEMENTS

Note: The author's acknowledgements have been reprinted from an earlier edition of this book. Sadly some of the people mentioned are no longer with us.

The author would like to thank the following for their assistance with information and illustrations for this book.

David Bate, Charles Cawood, Bill Collard, George Macdonald, John Melloy, Bob Moorhouse, Mervyn Spokes, Ron Stanbrook, Ford Motor Co. (Tractor Division), and those Fordson Major owners who took the extra trouble to return the forms with details of their E27Ns. George Walker formerly of Fryers of Uttoxeter and Jimmy Anderson formerly of Frews of Perth really had a lot to do with this book, as they were kind enough to dispose of all their redundant service literature in my direction.

THE FORDSON MAJOR

The largest manufacturer of tractors in the United Kingdom, the Ford Motor Company, had maintained a high level of production of its model 'N' throughout hostilities. The production line on which this model was produced was well nigh worn out by 1945, but it was not possible to retool and re-equip the plant to launch a completely new model. Firstly, the sophisticated machine tools required to cope with a new engine for such a tractor were not available from UK manufacturers, and spending in dollars was not encouraged. The continued need for tractors meant that the model could not be altered too much if production was to be maintained.

Consultation with the War Agricultural Committees identified the need for a tractor which could haul three furrows instead of two, have a rowcrop capability with higher ground clearance than the 'N', and have a central PTO. In addition the capability of the tractor taking a three point linkage and self starting were added by Fords, although these features did not appear until later. To get the ground clearance and greater power the old worm and wheel transmission of the Fordson were replaced by a new double reduction rear axle with spiral bevel drive on the bull pinion shaft, a feature copied from various US designs, such as the Massey Harris 203 and International "Farmall".. The result was the E27N (E = English; 27 = 27HP; N = Tractor). Some 60% of the parts used in initial production were the same as used by the previous model, a further 10% were N parts with modification, and the final 20% were all new, mainly in the area of the rear axle.

Unlike its predecessor which could be had also in Industrial and Pneumatic tyred (Land Utility) forms, it was available in two forms initially; Standard Agricultural, which had fixed track, steel wheels, and no independent brakes, and the Rowcrop, which had a beam type front axle adjustable for width, independent brakes, and steel wheels, the rear wheels being adjustable for width by juxtaposition of the hub and rim. The Land Utility, which had fixed track, brakes, and pneumatic tyres, did not arrive until mid 1945, nor did the rowcrop with pneumatic tyred option.

A certain amount of alteration was required to the assembly line to accommodate the new model, indeed both 'N' and 'E27N' tractors were built together for just under ten weeks. The major problem which arose was that the new units would not go through the old paint plant without causing problems, so the 'all over green finish' applied by a series of fixed jets mounted on hoops around the track gave way to hand held guns, and a new colour scheme - a return to the old blue and orange of the pre 1938 models. In fact the first E27Ns were taken off the line before painting and finished at the end of the truck line.

There was very little evidence of extras on the first twelve months production. Indeed the facility for self starting was there on the block, but it took until March 1946 before ring gears were fitted to the flywheel. At the same time the rather flimsy drawbar gave way to a more substantial affair. As is always the case, a number of modifications were evident after the initial production went into service and weaknesses were found. As patterns wore out further alterations were made to the design with the effect that this removed more and more of the pure 'N' features.

There is always an advantage in building a model which is a derivative of a previous one, and this manifests itself in the availability of spare parts for both the old tractor and the new. With much the same on the E27N it meant continuity of spares supply, and spare parts for the model it replaced were guaranteed, indeed certain what one might term exclusively E27N components found their way onto earlier models whether recommended for service or not. Magnetos, fuel taps, steering components, and improved design of governors and manifolds all found their way onto the 'N' in service.

In due course all the bolt on goodies became more freely available, and these of course included the simple PTO, the Hydraulic Power lift in two forms, the Smith Single Lever and the Varley Two Lever types, starting and lighting.

There was always one big disadvantage with the E27N, and that was the engine. It was outdated technologically, and this made overhaul a somewhat cumbersome operation. Unlike current Automotive practice, blocks were cast with cylinders en bloc; the connecting rods and main bearings were direct metalled. Quite sophisticated equipment had been developed at the dealers for the overhaul and machining of blocks, and oversize pistons were available, but on the whole the physical number of tractors in service by 1947 meant that dealers could not cope with overhauls, and thus a new scheme had to be instituted.

Thus it was that in 1949 that the Tractor reconditioned engine scheme was launched at the Smithfield show in December of that year. Initially a float of new blocks were launched of both N and E27N types and serviceable engines, less manifold, magneto bracket, cylinder front cover and plugs were returned to the factory for overhaul. Crankshafts were reground and blocks remettaled, and in general dry liners were fitted to the bores and all blocks brought back to standard. Engines of this type were identified by the tag on a cylinder head bolt which advised which type of crankshaft was fitted. In fact the supply of N blocks eventually dried up and it became necessary to use E27N blocks with N crankshafts and flywheels - hence the need for identification. The original engine number was ground off and a new T——-R number stamped on the LH clutch flange. The dealer was supposed to stamp the original engine (and chassis!) number on the "new" engine but many didn't bother, hence the problem in dating tractors nowadays. In retrospect it is a pity that the proposed development of the same engine to incorporate more modern features such as shell bearings did not take place, as it would have helped gain an edge on the competition!

It is all very well prolonging the life of tractors with an obsolete engine, but when the need arises to provide a diesel variant, it may sound strange that the engine manufacturer had more to do with this than the tractor builder. The Perkins P series had not been used widely in tractors until Frank Perkins converted a Fordson Major for his own use. The result was that Ford Motor Co. sent two Majors for conversion, one with a P4, and the other a P6. Both had fabricated sumps and flywheel housings. Whilst both gave a satisfactory performance, the P6 was chosen due to its lower maximum speed which suited the E27N gearbox better, and in adopting the engine as a production option it also must be remembered that the Ford dealer network were already geared up for Perkins spares and service as the engine in its vehicle (V) form was fitted in the Thames lorry.

So it came to pass that the P6(TA) engine was born and graced 23259 E27Ns, off the production line or an unspecified number as conversion packs which Perkins sold themselves.

233112 Fordson Majors were built from 1945-52, and a high proportion were exported, making 209853 units fitted with the Fordson s.i. engine. Headaches occurred for Ford especially in the form of competition from the "Grey Menace" and there were also other manufacturers on the scene by 1950, who although not capable of building as many tractors as Ford, with a good dealer and salesmen in a certain area, could knock holes in home sales.

Although Ford knew what sort of tractor if would have liked to build in 1945 it was not able to give the machine its proper engine. The E27N was a stopgap. That its basic transmission and rear axle survived until 1964 in the E1A range said something for the original design, weak though it was at first.

It is easy to date components on E27Ns, and we refer readers to the casting code key in the appendix. All major castings have a letter/figure/letter code set into the pattern at the time of casting which enables the exact date of casting to be ascertained.

Prototype E27Ns.

A number of people think that they have got the actual prototype!

I am sorry to tell you, but nobody has! If we look at the facts as they exist the first E27N was assembled on 19.3.45 and it was serial number 980520. This was the first tractor assembled on the production line, not the prototype. One would assume that casting codes on the first production tractors would be B25H to B28H and then C1H to around C15H, allowing for the time taken to cast parts and have them machined.

The prototypes on the other hand would be hand built by the development boys. One would imagine that castings used would be coded as most would be cast in Ford's own foundry. Indeed, apart from the rear and front transmission housings, axle housings, front axle, and a few other bits, the first few prototypes were built using mainly stock 'N' parts.

Only the prototypes had 'N' type radiators, which were, one would assume, later changed for the standard pattern. I would say that anyone with a 1945 Fordson Major should not get over excited unless the casting codes are before B25H!, and then it is possible that some prototype castings were not coded.

The E27N came in four variations. Firstly there was the Standard Agricultural with fixed front axle, fixed 9" width rear wheels and no independant brakes; secondly there was the Rowcrop, with adjustable front axle, independant brakes, and 4.5" wide rear wheels with detachable rims. By altering the position of the rim and the hub widths of track could be varied. The Land Utility proper, that is to say, a tractor fitted with pneumatic tyres in production, did not appear until June 1945, and the final model, the Industrial, till 1948. Two other variations, a tricycle/single front wheel model, and a low clearance model, were never put into production, and the crawler variants developed in association with County and Roadless (q.v.)

E27N production breakdown.

Built up:				Knocked Down:			
Year	Total	TVO	Diesel	Total	TVO	Diesel	Grand Total
1945	10412	10412					10412
1946	25290	25290					25290
1947	34915	34915					34915
1948	50561	50123	438				50561
1949	33213	27275	5938	162	162		33375
1950	35470	28528	6942	6805	6261	544	42275
1951	25679	17542	8137	10189	9257	932	35868
1952	416	88	328				416
Totals	215956	194173	21783	17156	15680	1476	233112

Fordson Major 1945-52.

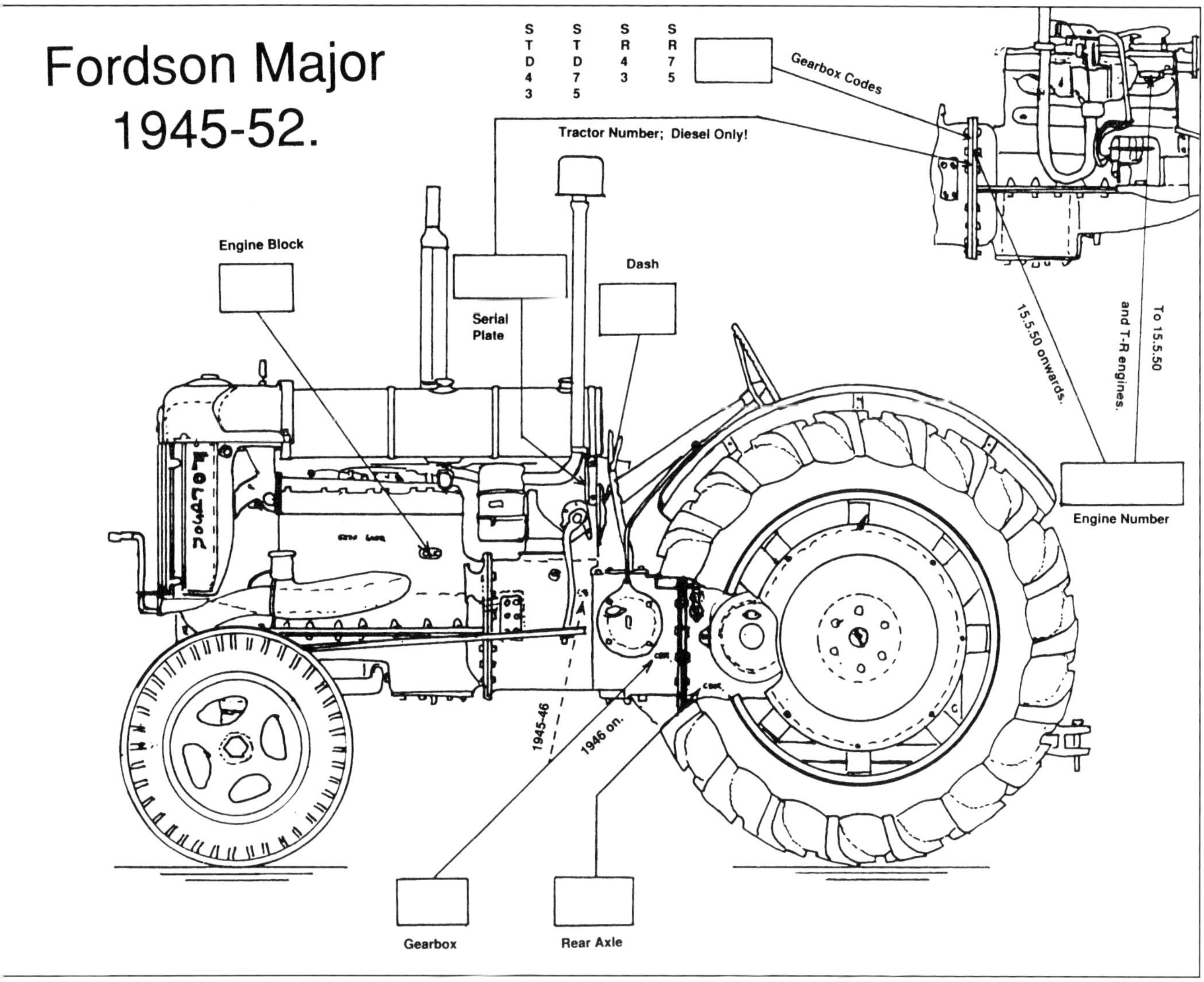

.Casting Codes. See diagram for main locations.
These are made up of three components letter/number/letter.
The first letter indicates the month.
The second letter indicates the date. (1-31)
The third letter indicates the year.

Year	Month
H = 1945	A = January
J = 1946	B = February
K = 1947	C = March
L = 1948	D = April
N = 1949	F = May
P = 1950	H = June
S = 1951	J = July
T = 1952	K = August
	L = September
	M = October
	N = November
	P = December

Example: D15H is 15th April 1946.

Fordson 'E27N'

Engine: Four cylinder side valve layout with cylinders cast en bloc. 4?” bore by 5" stroke giving 26.6HP@ 1200 RPM. Own Oilbath Aircleaner, or Burgess type from 1946. Own Vaporiser used with Low Compression Cylinder Head for running on V.O. or Zenith Carburettor used with High Compression Head for running on Petrol. Splash Lubrication. Lucas CJ4 Magneto, later 4 RF type.

Gearbox. Drive through single plate clutch to three forward and one reverse speed gearbox. Double reduction final drive with differential on countershaft and slewing brakes on outer ends of bull pinion shafts.
Power Take off for HPL on top of gearbox, for rear PTO on bottom.

Colour. Dark Blue with Orange Wheels. (a few early examples green)

Weight 2 Tons..

Top. The E27N as originally envisaged, in section. Note the artist's use of the N style radiator.

Details of early tractors:

A: Attaching extension rims to the 4½" wide rear wheels of the rowcrop tractor.

B: The vaporiser, governor and magneto; note the control rod passing through the manifold shield.

C: The rear axle jacked up also showing the location of the brakes.

D: The location of the gear selectors and brake pedals.

E: The rear of the tractor showing original type drawbar and central PTO installation.

F: The handbrake, which operated the transmission brake inside the gearbox, showing the interlock with the clutch pedal.

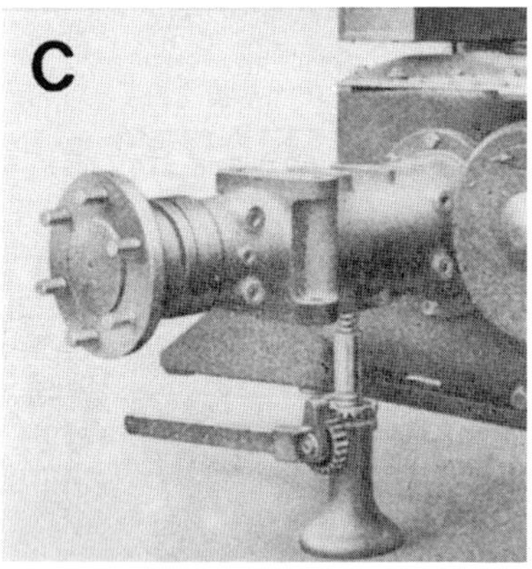

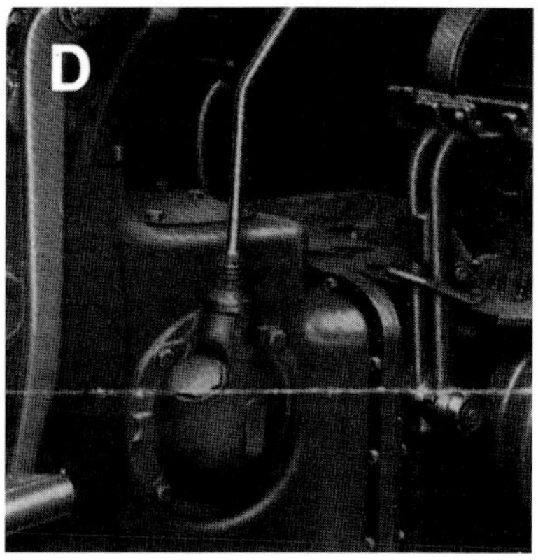

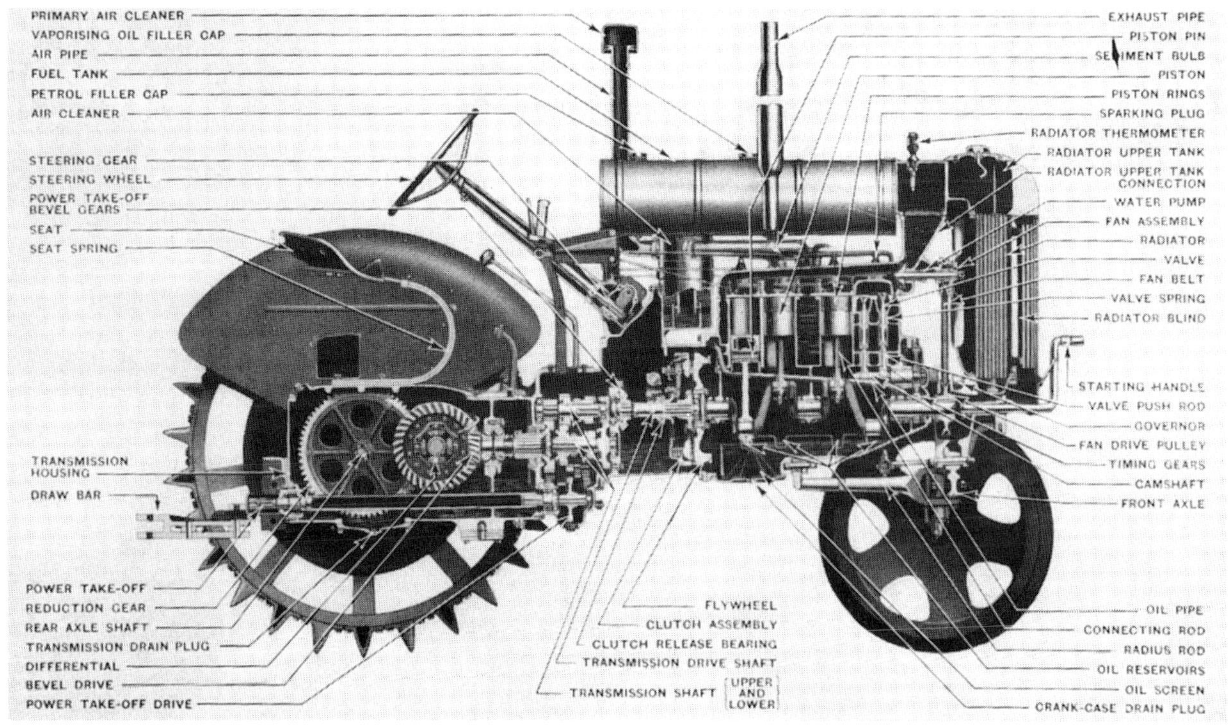

The revised sectional drawing shown in later literature.

These views show artists retouchings of one of the prototype tractors used in early sales and service literature. Note the non standard rear wheels, the shape of the radiator top tank, and the mudguards. The two lower illustrations offer direct comparison between Standard Agricultural (left) and Row Crop (right) models.

Three views of early rowcrop tractors. Note the first type of mudguard used in production, the offset governor control rod, the pillar calormeter, square aircleaner, and N type exhaust.

This illustration shows a later Standard Agricultural tractor with later type drawbar, 'Burgess' aircleaner and exhaust. Note the cones covering the bull pinion shafts. The rear wheels were 9" wide.

This later 1946 tractor has the second pattern of mudguards. Note the continued use of the 'N' lower dash casting with its mounting for the earlier type of square aircleaner. This tractor would have cost £226 in the form shown.

The Rowcrop featured an adjustable front axle, independant brakes and 4½" wide rear wheels which could be adjusted in width from 54" to 72". This early 1946 example would have cost £285 in basic form.

***Top, left and bottom:* The first Land Utility ever built on the assembly line was KHK539 which found its way onto the Henry Ford Institute's farms at Boreham. 9.00 x 36" rear tyres were used at this time.**

***Below:* A number of features of E27Ns built from 1946 on are seen here including the governor control rod outside the dash. A pulley is also installed on this example.**

***Above and Right:** Wheel grip was a problem with early pneumatic tyred tractors, and the example seen here was fitted with 'Case' type Dunlop cast centres and tested at the NIAE at Askham Bryan near York. In the end, the grip problem was solved by the raising of the front axle stool slightly (something which was never advertised! This moved the centre of grip on the rear wheels at right angles to the ground. The final solution was, however, to use 11" rear tyres.*

***Bottom:** It was 1946 before hydraulic power lifts came in. This tractor is obviously a Land Utility (note the Lucas No. 12 bulb horn) equipped with spade lugs for ploughing. This practice was common on heavy land right up to the seventies when 4WD and other means of grip enhancement were taking over.*

This 1946 Land Utility clear shows that the brake peda missing from the Agricultur model are clearly evident as the use of the AC fuel changeov with glass sediment bowl. Th steering drop arm is the typ supplied when mid or fron mounted implements were use these seemed to be more commo on 1946 built tractors possib due to a shortage of the othe type, or the fact that large stock of the cranked pattern we ordered and not required tha much in service. The LU. co £281 in 1947 - all such tractor had rear view mirrors and Luca No. 12 Bulb horns - but where i the horn in this picture?

A late 1948 Land Utility tractor complete with starting, lighting and Hydraulic Power Lift of Smith type. The last type of mudguards are now fitted, and the later pattern of flat based lower dash. The first type of indirect starter is fitted, with separate solenoid, and the small semi circular instrument panel is attached to the steering column.

Bottom: A display of E27N tractors and equipment in the Shifnal works of Reginald Tildesley Ltd on the opening day in 1949. There is much of interest in this illustration, including the engine block upturned on the bench in the centre foreground about to be line reamed after remetalling.

Above: The Industrial was offered from 1947, with 16"brakes on the rear axle shafts, windscreen, automatic drawbar coupling, plus the canvas canopy which was optional. Note also the direct throttle, the tapping off the manifold for the vacuum operated windscreen wiper, the underswept exhaust, and the Vokes type dry air cleaner.

Below: A bit of careful retouching in the studio has produced this illustration of a 1947 Land Utility as fitted with 14 x 28"rear tyres and 7.50 x 20"front tyres, to suit local conditions, for sale in Australia. The wheels were made in Australia, there being restrictions on the importation of rubber goods into this province at the time.

Two tractors were sent to Perkins at Peterborough to have diesel engines fitted. One was equipped with a P4(TA) engine as shown here. All parts were fabricated, note the space between the radiator and the timing cover. The tank has also been shortened slightly. Whilst this conversion worked well, for ideal conditions it would have needed a change to either the gearbox or rear axle ratios to deal with the full power output of the engine.

The P6 conversion was chosen as the engine fitted more neatly and could develop the required power using the existing gearbox and rear axle ratios. The electrical arrangements here are of course non standard, and the sump, front mounting, and flywheel housings fabricated.

Late 1948 saw the P6(TA) engine offered as a factory fitted option. An early example is seen here with original style of pre-cleaner (4 flat vanes). Note other contemporary features such as the last type of mudguards with toolbox mounted, as it was on all diesels, on the lh mudguard, and the addition of 300 lb. of wheel weights to the rear wheels. Front wheels now have the dirt ring cast inside, and the front grille is the narrow bar type.

The greater battery capacity required to start the P6(TA), necessitated a revised electrical system which used 2 six volt batteries. A Ki-Gass pump was provided to aid starting. This view also shows how neatly the P6(TA) engine fitted the E27N.

A 1950 example is shown here, b this time the P6(TA) engine ha been improved somewhat an revised breathing arrangement provided. The large domed pre cleaner also appeared in earl 1949 due to the flimsiness of th earlier design. The use of standard VO pre-cleaner had detrimental effect on the P engine's pneumatic governor.

The P6(TA) engine as fitted to the E27N was the 'low speed' variant of the engine designed to run up to 1500RPM and featured cast iron pistons. At maximum governed speed it gave 46BHP.

Perkins engined E27Ns for the Australian market had the usual wheels fitted there, plus Yuba spark arrester exhausts. Note the fitting of the "37-41 diesel" badge on the radiator grille, and the linkage for the radiator shutter.

Space precludes showing all but the most common of E27N variants in this title.

Above and Below: DG4 tracks were based on the principle of having a driving wheel equivalent to one of some 20 feet in diameter by virtue of the tracks being mounted in tension to hold the profile to the ground in a large arc. Roadless traction supplied these tracks for both TVO and diesel tractors as seen here.

Skeleton DG4 tracks were also produced. This 1949 tractor so equipped also gives a good insight into the other features common on E27Ns at the time.

Left: Another view of the DG4 skeleton equipped Major. The lack of air intake stack was a common 'in service' modification to the E27N. Owners (and even dealers) removed the stack and fitted the pre-cleaner straight into the main aircleaner body.

Above: This rear view of a P6(TA) engined Major with DG4 tracks shows certain features unique to the diesel model such as the location of the toolbox, the battery layout, and the control box location. Other points of note are the tail lamp and socket, and the mounting of the idler sprockets and arrangements for keeping them in tension. It was intended that the DG4 tracks could be removed in the Summer and wheels substituted, but this was not often done. In addition, the Hydraulic Power Lift equipment could not be used with DG4 tracks.

The County Crawler used most of th E27N transmission but the differentia was replaced by a solid drive and clutch brake units were mounted on the bul pinion housings with concentric inpu and output. The tracks had to be of rathe large diameter, but the crawler, built a Fleet, Hants., sold well enough. Diese engined models were the most popular and most of the E27N extras could be used Advertised as a 'proper crawler' this uni was really a conversion, and could be ha in 12", 14"(left) and 16" (below) trac widths.

Bottom: Roadless took some time before their type E tracks were applied to the E27N. This prototype used steering arrangements similar to the County, but to give better weight distribution the unit was extended by the insertion of a spacer between engine and gearbox. The tracks were of the usual rubber jointed variety.

Left: A narrow version of the E27N with DG4 tracks was also produced for used in North Africa.

Right: A number of production full crawlers were built by Roadless but only accounted for around two dozen units. The tracks on the production version were longer and had four bottom rollers.

This E27N with equal sized 11x38" front and rear wheels was built by Roadless for use in the swamps of Tanganyika. To aid steering a larger diameter steering wheel was provided.

Above: The penultimate in E27N production is exemplified by this May 1950 example of a Rowcrop tractor. The later type of electrical arrangements are provided with direct starter motor; the revised instrument panel can be seen clearly. Also fitted is the improved Vaporiser.

Of the extras fitted, lighting and starting added £20, PTO £11, Pulley £12, Hydraulic Power Lift £38, and linkage £21 to the basic price of the tractor.

Below: The final Australian model is seen below. The 'Yuba' spark arrester, Major decals on the tank, and wheel equipment all show a model assembled in Australia from CKD form for that market place.

***Above:** The E27N which was seen by millions! Mounted on a hydraulic ram at the 1951 Festival of Britain on the South Bank in London, with electric motors rotating the wheels. This tractor shows the final appearance of the E27N before it passed into history!*

***Below:** Perkins introduced their L4(TA) engine in 1953, and produced a conversion pack for the E27N. A large number of these engines and the later four 270 engine were fitted into E27Ns; of course this engine was never used in production.*

We now take a closer look at the tractor, component by component. The order in which this is being tackled is in line with the Ford universal part numbering system.

Front Axle and Front Wheels.

From the start two front axles were available, the fixed variety on the Standard Agricultural and Land Utility variants, and the adjustable on the Rowcrop. The front axle was a heavy steel casting, in the case of the Standard Agricultural, Land Utility and Industrial models a single piece casting, and in the case of the Row Crop a single centre beam was bolted to two outer beams which could be repositioned to alter the track from 48" to 72". Unlike the Fordson, which featured plain steel kingpins retained in bronze backed bushes, the E27N front wheeel spindles were retained in a different manner, in order to give maximum axle clearance, and incorporating a ball type thrust bearing. The original arrangement caused some problems, and in fact was modified 5 times from 1945-49.

The front wheels were carried on similar taper roller bearings to the previous model. The steel wheel and pneumatic centres were, in fact, of similar design, though not identical, to the units fitted to the model 'N'. The pneumatic tyre size was of course 6.00 x 19". In 1948 the wheel centre was redesigned to incorporate a dust ring.

The rear wheels were now retained on the rear axle shaft by six studs and nuts. The Standard Agricultural had steel rimmed rears 9" wide mounting 24 spade lugs. The Land Utility had Sankey type pressings mounting 9.00 x 36" tyres on rear with 10 locating lugs; later 11.00 x 36" rims with only eight locating lugs were fitted. The Rowcrop had rear wheels 4½" wide mounting spade lugs or Sankey type pressings mounting 9.00x36" tyres with width adjustable by juxtaposition of rim and centre. Again 11.00 x 36" rims were universal from 1947.

Tractors supplied to Australia were shipped out initially fully assembled either on steel wheels or less wheels. Where pneumatic tyres were fitted these were of different sizes to those used in the United Kingdom, and had distinctive centres. The normal sizes were: 7.50 x 20" on the front and 14 x 28" on the rear but other options to suit local conditions were available.

The Rowcrop axle is seen, complete, and in section, on this page. Whilst the assembly looked all new, in fact a number of Fordson 'N' parts were still used. Early tractors did not have the cross piece on the radius rod, and the sides of this were partly in I section. Adjustment was effected by jacking up the axle and undoing the six bolts; the disposition of the holes being such that track could be increased in 4" increments. It was also necessary to slacken off the track rod clamps.

Considerable trouble was experienced with the early front axle spindles, and it took some time to rectify this problem. The appended table shows the modifications applied. Standard Agricultural, Land Utility, and Industrial tractors were fitted with the non adjustable axle shown (opposite top).

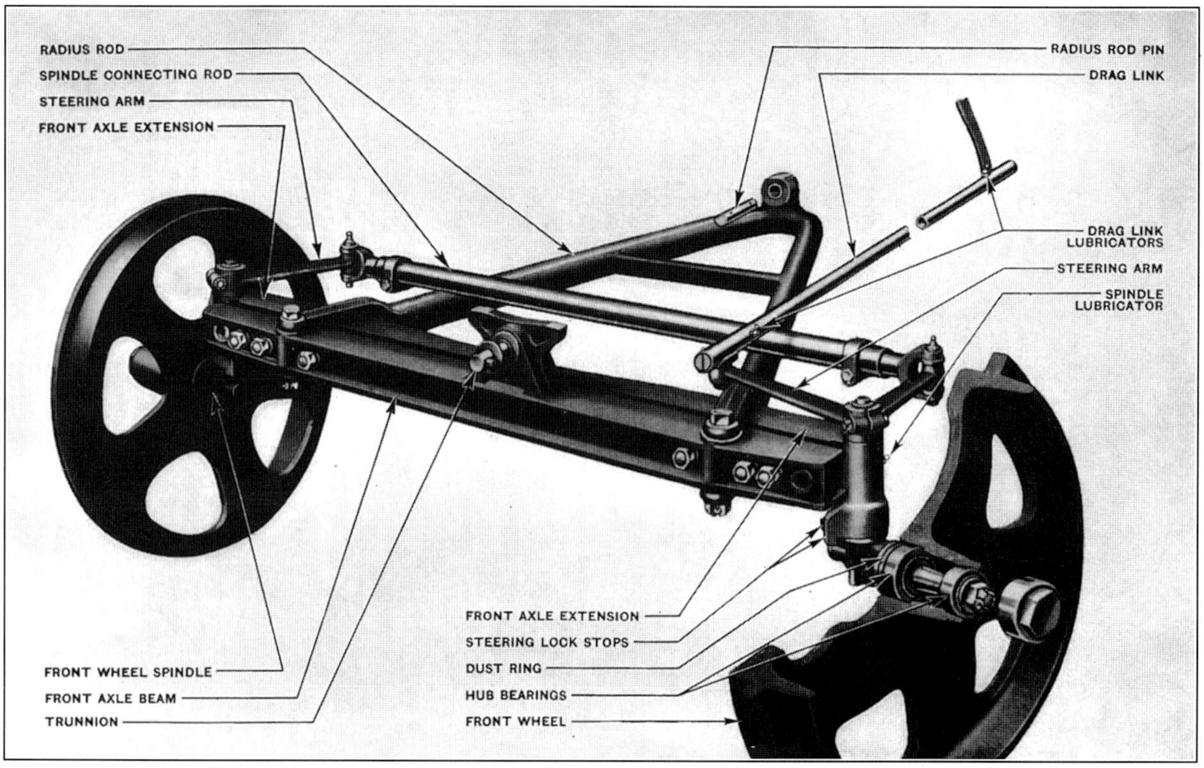

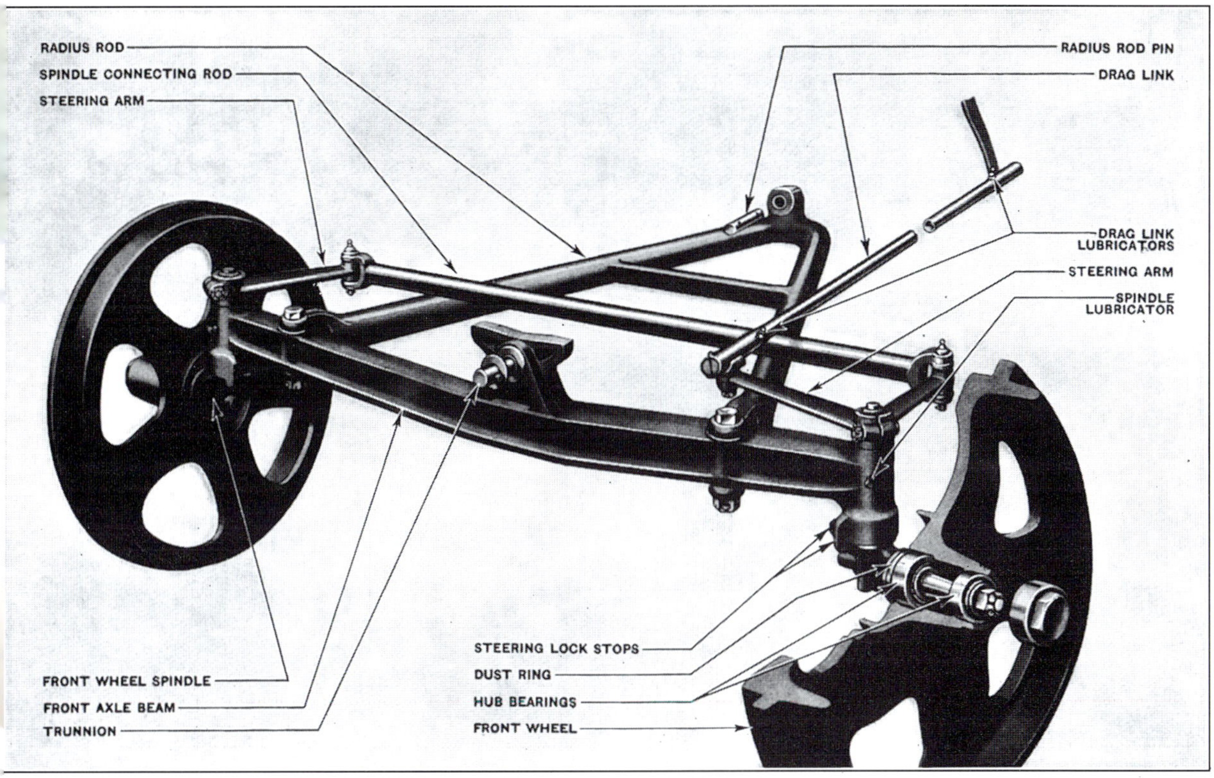

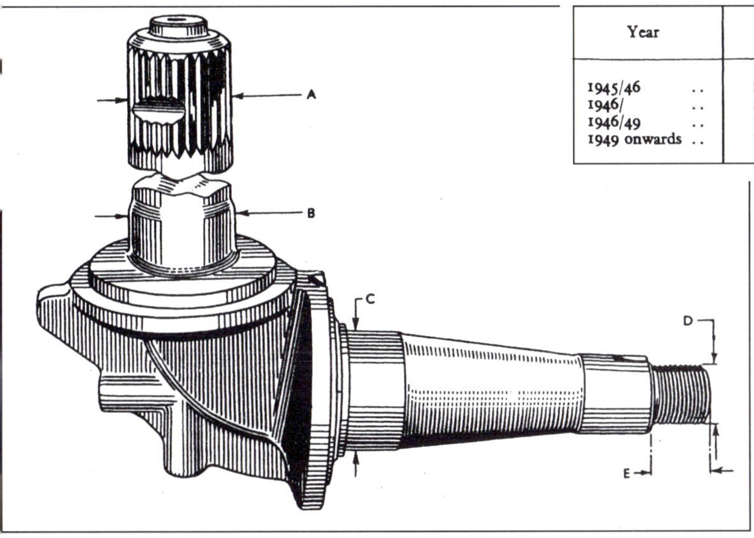

Year	A	B	C	D	E	Spindle Nut Part Number
1945/46 ..	$1\frac{3}{8}$"	$1\frac{1}{2}$"	$1\frac{5}{8}$"	$\frac{3}{4}$"	1.09"	351125–S
1946/ ..	$1\frac{3}{8}$"	$1\frac{1}{2}$"	$1\frac{5}{8}$"	$\frac{7}{8}$"	1.09"	351152–S
1946/49 ..	$1\frac{1}{8}$"	$1\frac{5}{8}$"	$1\frac{3}{4}$"	$\frac{7}{8}$"	1.09"	351152–S
1949 onwards ..	$1\frac{1}{2}$"	$1\frac{5}{8}$"	$1\frac{3}{4}$"	$\frac{7}{8}$"	1.25"	E27N–3161–B

***Left and Above:** This diagram and table shows the changes to the front wheel spindles on the E27N over its production run. **Below:** The components of the fixed front axle showing the later type radius rod.*

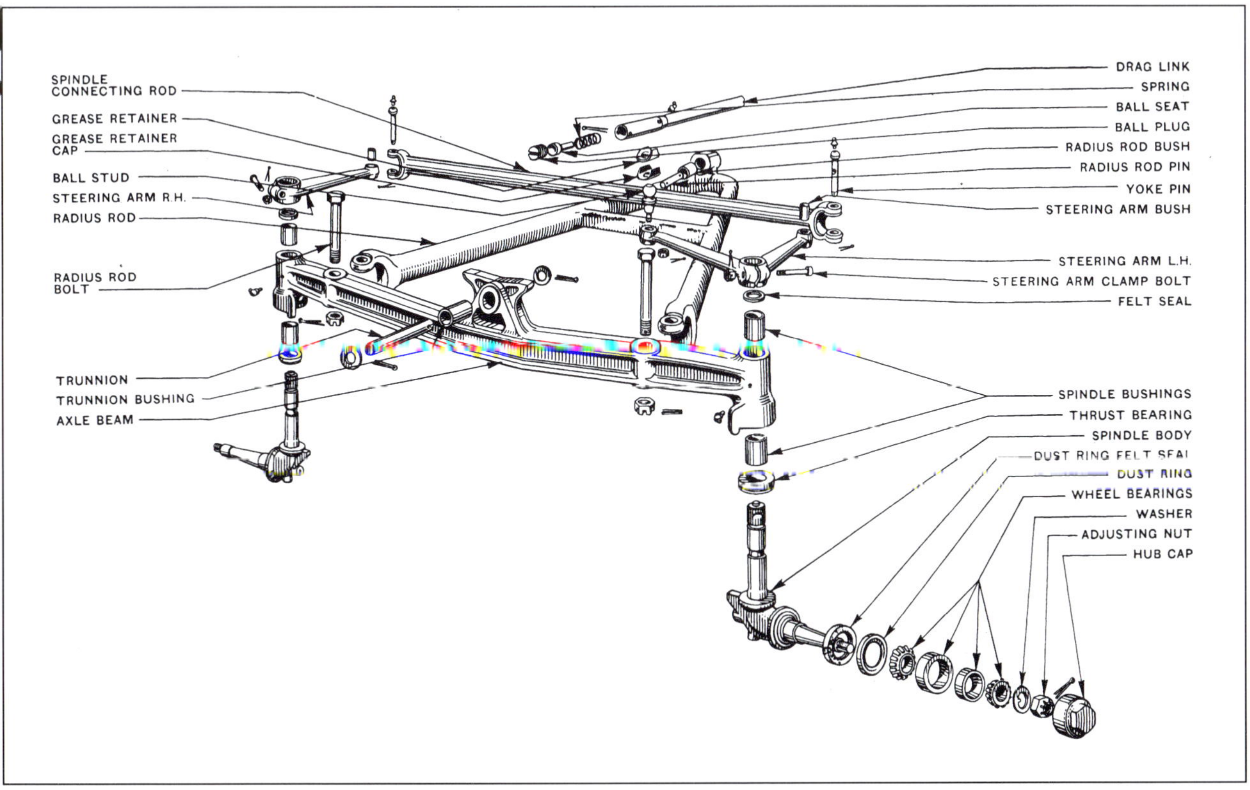

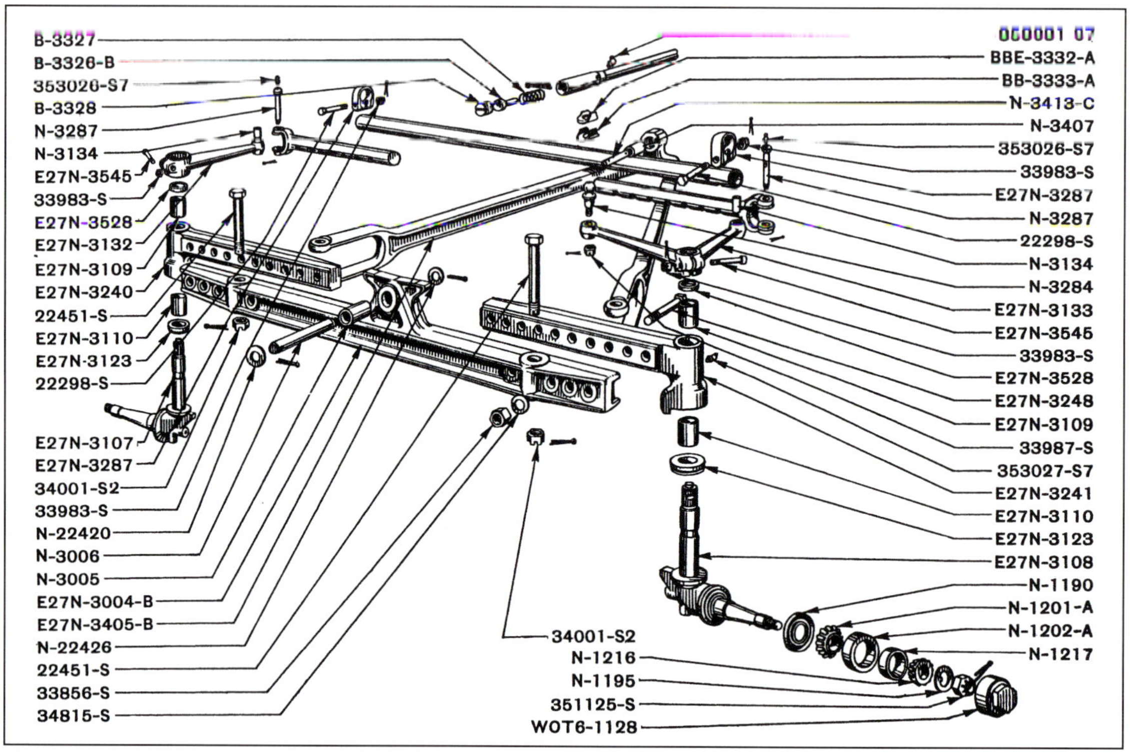

The adjustable front axle exploded. This drawing shows the original type of radius rod used in 1945.

Brakes

The Land Utility and Rowcrop were fitted with independent steering brakes on the ends of the bull pinion shafts. These were self energising, cable operated units, the foot pedals being mounted on the left side of the transmission housing, and capable of being locked together for use on the road, and locked down to provide a parking brake. The advantage of using these dual shoe brakes on the countershaft rather than the rear axle is explained by the fact that the bull pinion shafts revolve five times faster than the rear axle shaft, and the use of almost standard automotive components resulted, and the added advantage that the leverage provided positive 'locking' when using the brakes for steering assistance. Until September 1948 the covers were attached by three lugs to brackets on the backplates, from that date setscrews were used through the rim of the cover into the wider rim of a modified backplate. Oil leakage from the bull pinion shafts onto the brakes was something which was never completely cured, hence the extensions adopted on the New Fordson Major.

The Industrial tractor used brakes fitted onto the rear axle shafts; a single pedal was provided.

Transmission Brake

The gearbox contained the same transmission brake as the Fordson 'N', but its operation was transferred to the rear of the assembly from a lever operated cam. The handbrake lever was mounted on the right side of the transmission housing and was so designed as to prevent operation with the clutch engaged. Two types of handbrake lever were used, a forged design and a pressed steel design, the latter not becoming common until 1948.

Steering Gear

Until 1947 the previous design of Fordson Worm and Sector steering unit was mounted on the lower dash. The steering column on all but the earliest tractors was longer however. From 1947 a new worm and nut steering assembly was introduced, from tractor serial number 1022878. Two types of drop arm were in use, one straight type supplied as standard, and a cranked type fitted where midship or front toolbars were used. For a period in 1946 all tractors were fitted with the latter type, no doubt due to a shortage of the former. The drag link was the same type as used on the Fordson with spring loaded ends. Early tractors up to 1947 had the same bakelite covered steering wheel as the 'N'; some tin steering wheels were also used from 1947-9. A much neater alloy steering wheel came in 1949. A larger diameter steering wheel was fitted to some export machines.

Above & Right: The original type of steering brake and cover are seen here.

Below: These components were modified from September 1948 and the revised design of brake plate, shoes, and cover is seen below.

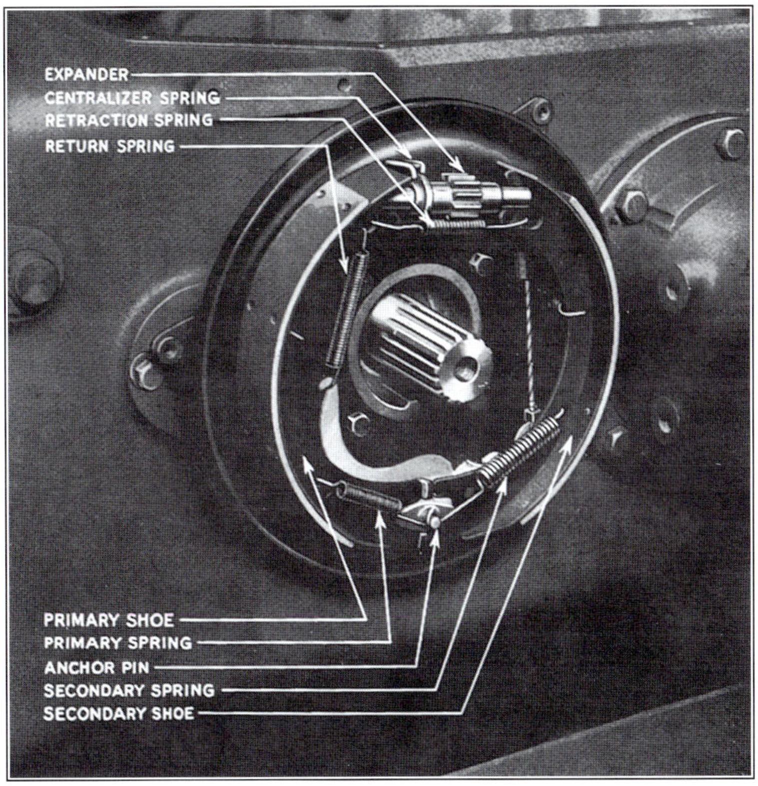

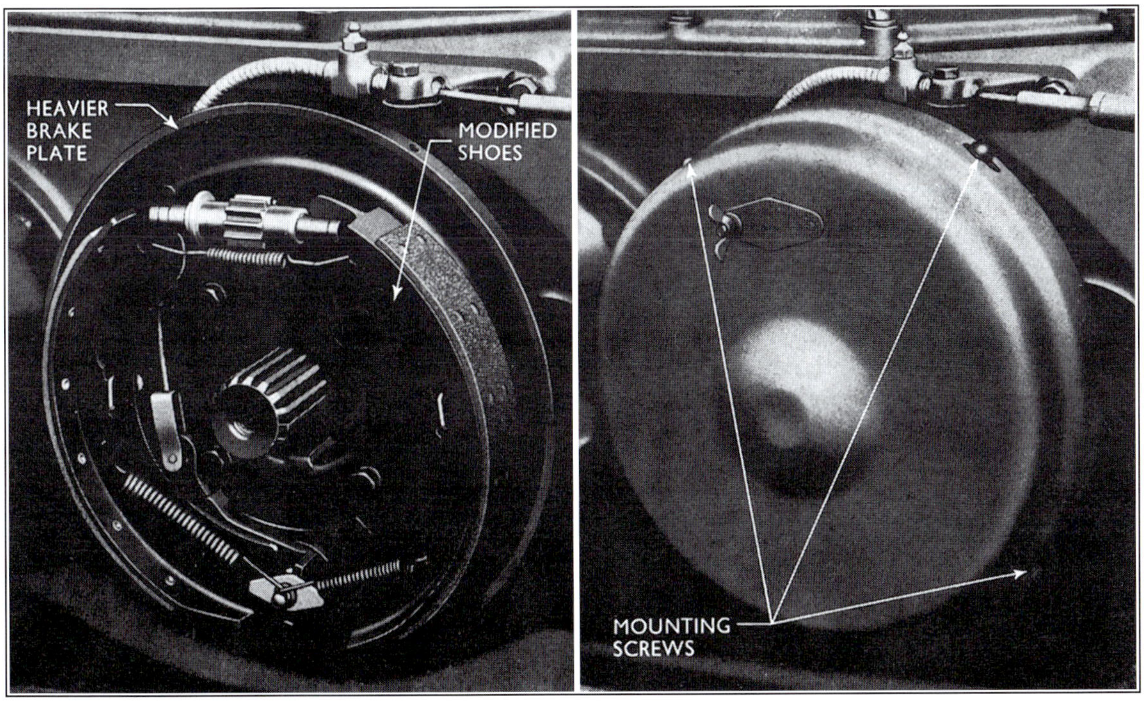

Fordson Major Serial Numbers.

First 'E27N' built 980520 on 19-3-45

1945	980520	1949	1104657
1946	993489	1950	1138235
1947	1018978	1951	1180610
1948	1054094	1952	1216575

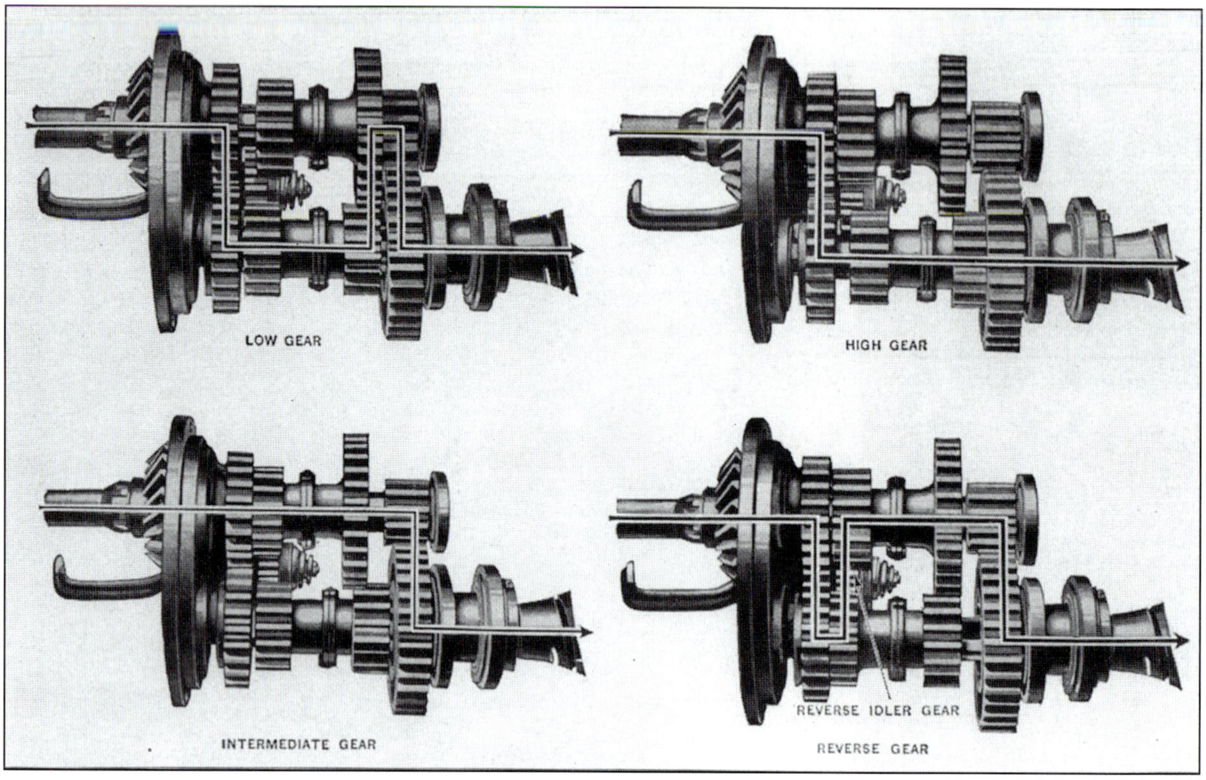

Gearbox

The gearbox used in the E27N was in fact identical to that of its predecessor and the same ratios were available (see table). The input shaft was altered to suit the new clutch, and the hook used to operate the transmission brake on the engine side of the transmission plate was removed, this being operated by a cam from the rear. The gear selector assembly was also identical to that used on the 'N', but the gear lever was straight for some distance before bending rearward. Tractors with Perkins diesels had a further bend in the gear lever to clear the battery carrier.

Above: ***Gear ratios are seen here. Although the illustration shows a model 'N' gearbox the E27N equipment was virtually identical. A 7.7 transmission is seen here.***

Below: ***An exploded view of the gearbox housing and PTO.***

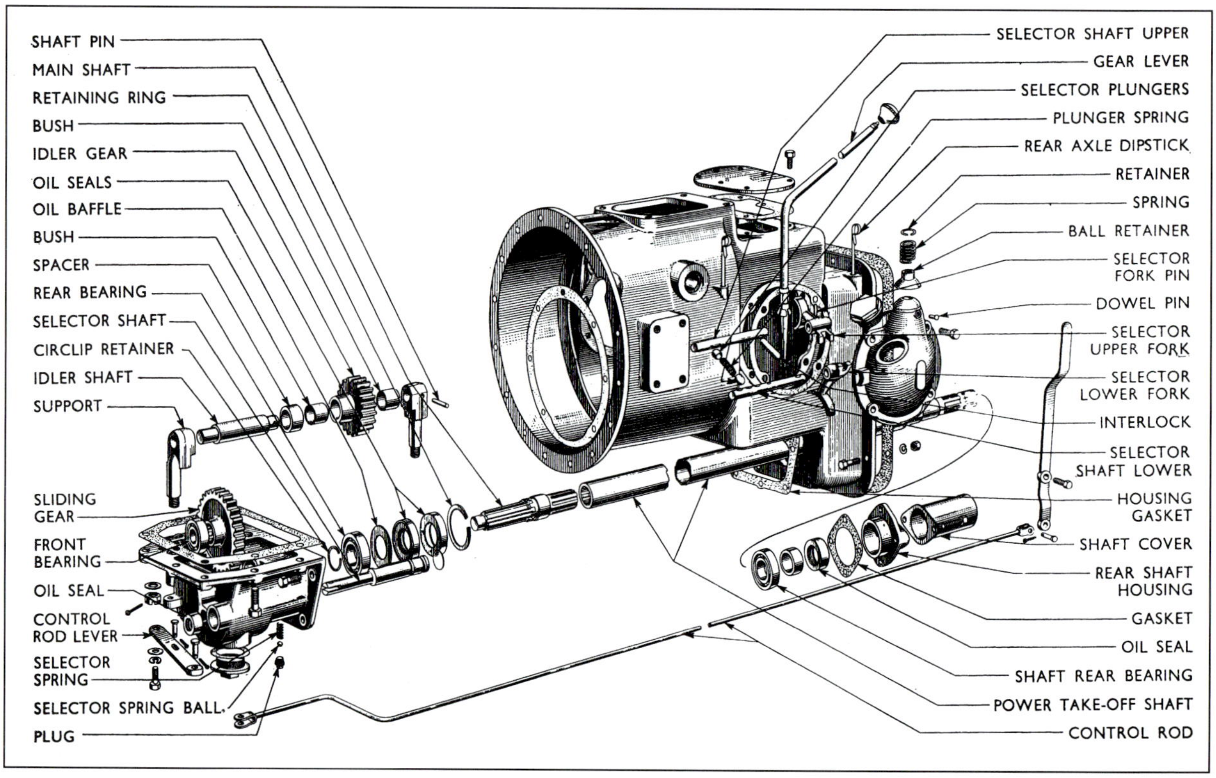

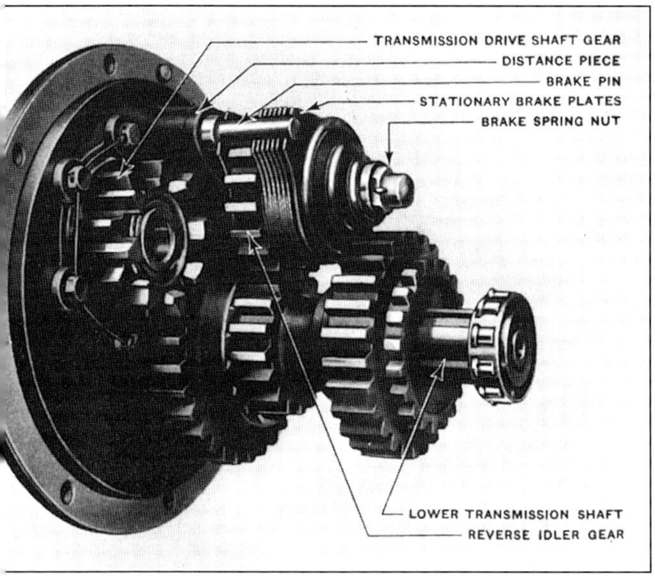

Above: ***The 4.3 transmission with the transmission plate and its associated gears withdrawn from the tractor. Note the castellated gear on the drive shaft held in place by wired setscrews, an identifying feature of this gearbox.***

GEARBOX RATIOS.	1st	2nd	3rd	Rev.
Standard Green Spot	2.35	3.31	8.30	3.18
Special Red Spot	1.86	2.62	8.30	2.48
Standard Green Spot with 4.3 Transmission.	2.18	3.31	4.61	1.88
Special Red Spot with 4.3 Transmission.	1.72	2.62	4.61	1.48

Rear Axle

The rear axle was of course the one item which was completely new on the E27N. The original housing lacked the implement mounting pad on the rear, and was of lighter construction. A number of failures caused the redesign of this component with thicker section, fitted with the additional Implement mounting pad at the rear, and used from February 1946. A special housing was provided from 1947 to suit the industrial tractor to allow the speedometer drive to be fitted. Two types of rear axle housings were therefore also used, those able to take brakes for the Industrial, and the plain ones used on all the other models. When brakes were not fitted on Agricultural models cone shaped covers bolted over the ends of the bull pinion shafts. The original rear axle housing cover was also found to be inadequate when hydraulics were fitted, and an improved design was fitted from the end of 1946, from serial number 1016085. This was easily identified by the letter H embossed on the cover. The earlier pattern of cover was also modified when the width of the final reduction gear was increased to 1.75". The original rear axle shafts had a separate stud screwed in to the end to hold the final reduction gear in place, but one piece axle shafts came in March 1946. Finally, the introduction of the diesel engine saw problems in the area of snapped half shafts, and this saw the evolution of a new type shaft with 19 instead of 10 splines, and 2.5" in diameter rather than 2.25". These were identified by having first 2 1/2 and later "I" stamped on the outer end of the shaft. Tractors from 1141433 had the reinforced shafts.

Power Take Off

The provision of a central PTO was achieved by taking the drive from the bottom of the gearbox and making the rear axle housing lower than the gearbox. Either an L shaped blanking plate was fitted to tractors less PTO or the drive unit, which required different components to suit high or low top geared tractors. The Hydraulic Power Lift drive came off the top of the gearbox and the shape of this unit was identical with the side PTO fitted to the Fordson N with 4.3 transmission but with an extra piece cast on one side. Again different gears were required to suit high or low top gear transmissions.

Drawbar

The original pattern of drawbar fitted to the E27N only provided for lateral adjustment and was fitted up to tractor 996902 in February 1946. When the most substantial unit with vertical adjustment was introduced (at the same time as the heavier rear transmission housing with implement mounting pads) it only had three positions. In December 1946 a new rear frame was introduced with four vertical positions. In 1949 the drawbar frame guide assembly was modified and attached by stronger bolts. The Industrial tractor had a sub frame to which was attached the usual WD pattern automatic hitch.

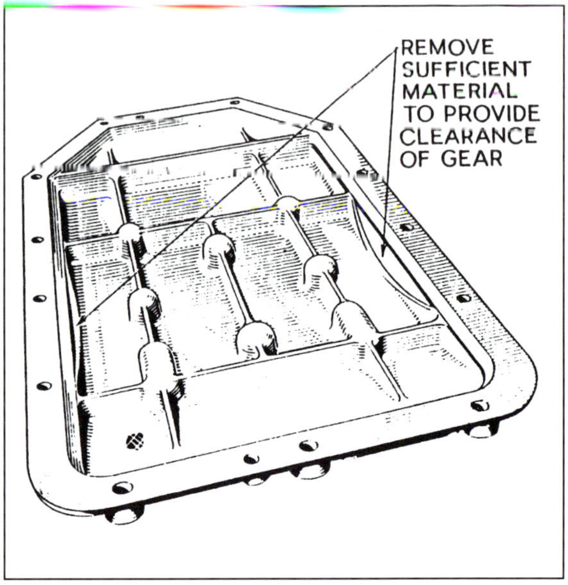

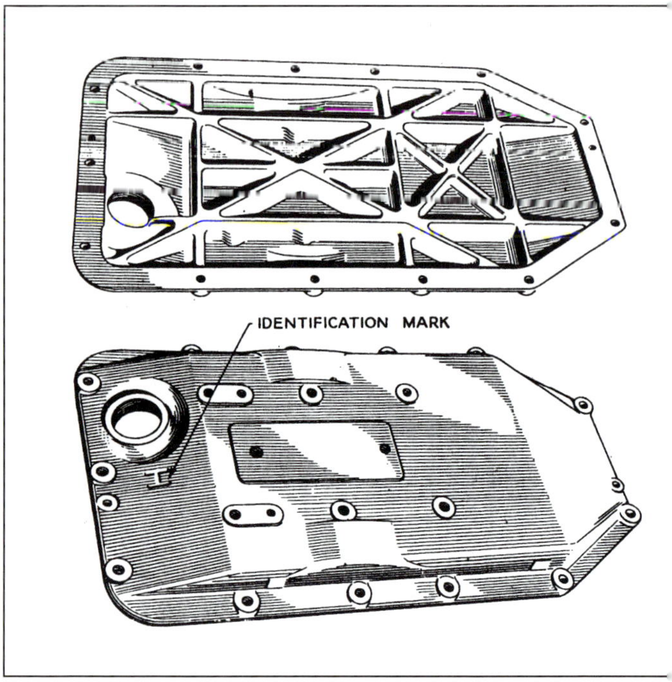

***Top right:** The original pattern of transmission cover.*

***Above:** In order to carry the loads imposed on the rear axl housing by the hydraulic power lift a reinforced housin was fitted from late 1946.*

***Left:** The transmission line of the E27N showing clearly th differential cage and the large reduction gears.*

***Below:** The Gearbox and rear axle in section.*

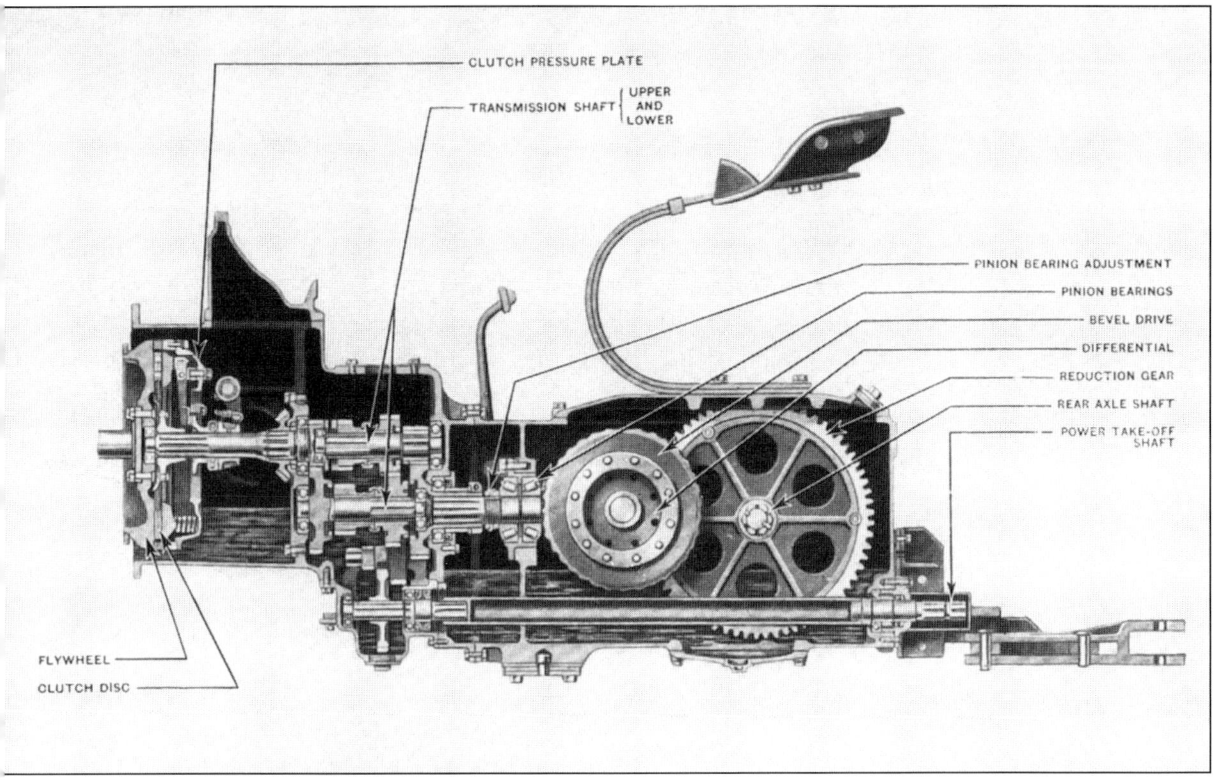

Above: The rear axle and gearbox in cross-section. Note the respective oil levels in engine, gearbox and rear axle, necessitating the oil seals between them being in good condition.

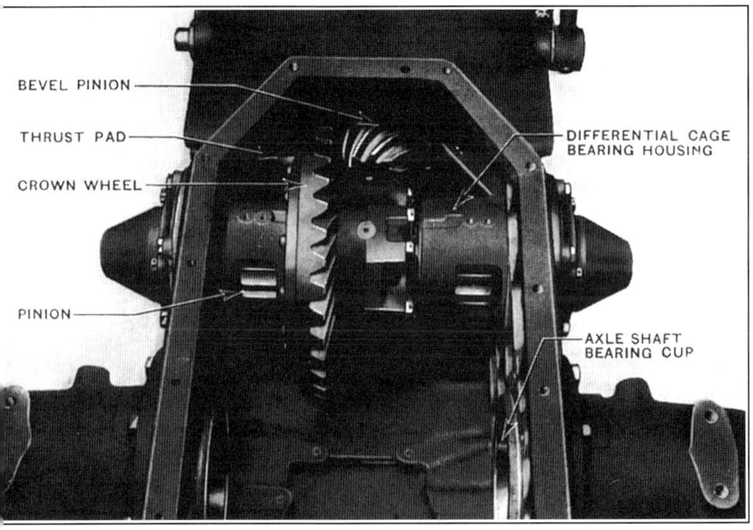

Left: The differential in situ. Note the coned covers used to cover the bull pinion shaft ends on Standard Agricultural and Industrial tractors.

Below: An exploded view of the internal components of the rear axle.

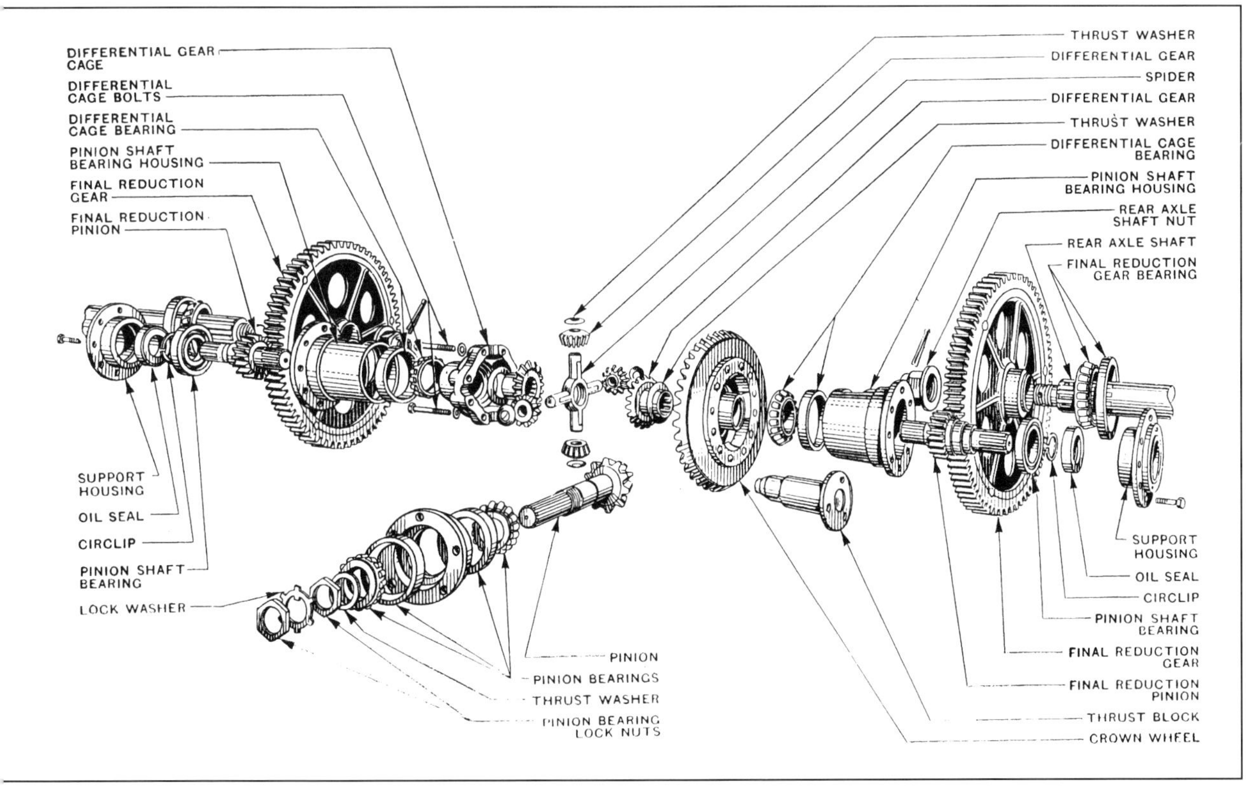

Left: The engine out of a 1948 tractor. The engine was to a design dating back to 1917. The original type Vaporiser, Lucas 4RF magneto, and four bladed fan are seen here.

Engine.

The engine used in the E27N was originally no different from the model 'N' unit apart from the provision on the block of a starter motor aperture on the nearside. Three variants were produced, the Vaporising Oil unit with cylinder head N-6050-E; petrol with head N-6050-C - both with flat pistons; and Gas Oil with head N-6050-B with domed pistons. For service pistons and rings up to 50 thou oversize were supplied, as were dry cylinder liners to put the blocks back to standard. One feature which was deleted from E27N blocks was the provision of core plugs in the front and rear of the block. The flywheel was of course designed to suit the single plate clutch of the E27N, and early tractors did not have a starter ring gear fitted; it was not a standard fixture until 1st March 1946. An internal modification to the engine from 1011911 concerned the oil pipe. It was foun[d] that the fitting of the starter ring gear disturbed th[e] oil flow into the end of the oil pipe and so this wa[s] redesigned with a deflector rib fitted to the clutc[h] housing, allowing the lifting effect of the ring gear t[o] be used to advantage.

The cylinder front cover used on the E27N wa[s] based on that used with the straight axle on th[e] Fordson 'N' to allow attachment of tool bars etc.

The diesel engined variant used the Perkins P6(TA[)] engine, details of which are outwith the scope of thi[s] 'Special', but these can be found in 'Special 8 - Perkin[s] Diesel Conversions and Factory Fitted Units.

One feature of E27N blocks was the early deletio[n] of the core plugs in the ends of the block, leaving onl[y] the three underneath the manifold.

Left: A late engine with a difference. This example is out of an Australian tractor and although it features the 1950 Vaporiser, the double belt drive for the dynamo and extended manifold drain valves are unique to this territory.

Opposite page. Top: Every tractor reconditioned engine was delivered with a small orange & black envelope label attached in which were provided the instructions reproduced alongside. Handling of such units by dealers often became quite sophisticated as seen with the vehicle here operated by Reginald Tildesley of Willenhall. The label can clearly be seen tied to the front of the cylinder head.

TRACTOR RECONDITIONED ENGINE

ITTING INSTRUCTIONS

1) The following items are to be removed from the old engine and fitted to the reconditioned engine :—
   - (a) Sparking Plugs.
   - (b) Clutch (in the case of Fordson Major Tractors).
   - (c) Fan and Water Pump.
   - (d) Starter Motor and Generator with mounting brackets (when fitted).
   - (e) Vaporiser Assembly, together with inlet and exhaust manifolds and studs.
   - (f) Governor and Controls.
   - (g) Crankshaft Pulley and Cylinder Front Cover.
   - (h) Magneto and Bracket Assembly, together with H.T. wiring and conduits.
   - (i) Oil Filler and Breather Pipe Assembly.
   - (j) Tool Box (when fitted).
   - (k) Any other engine accessories.
2) Drain off all old oil and wash down the engine.
3) Remove the reconditioned engine from its special crate. Remove the various aperture covers from the cylinder head outlet, cylinder front cover, inlet and exhaust manifolds, etc. Remove the dummy sparking plugs from the cylinder head. Transfer these covers and plugs to the old engine.
4) Pack the old engine unit in the special crate and return it to the factory, carriage forward.
5) Refit the components removed from the old engine to the reconditioned engine. Ensure that the magneto is correctly timed, and check the sparking plugs before refitting them.

ERVICE NOTES

When fitting the engine to the Fordson Major Tractor transmission, ensure that the oil pipe fitted is of the same type as that itted to the old engine.

When fitting the starter motor to the Fordson Major engine unit, use the composition-type (not cork) gasket. Tighten up the ecuring screws evenly to ensure correct alignment of the starter notor.

Tighten the crankcase bolts, left loose for ease of fitting timing over to engine.

Stamp original Engine No. on the new engine.

ORD MOTOR COMPANY LIMITED, DAGENHAM

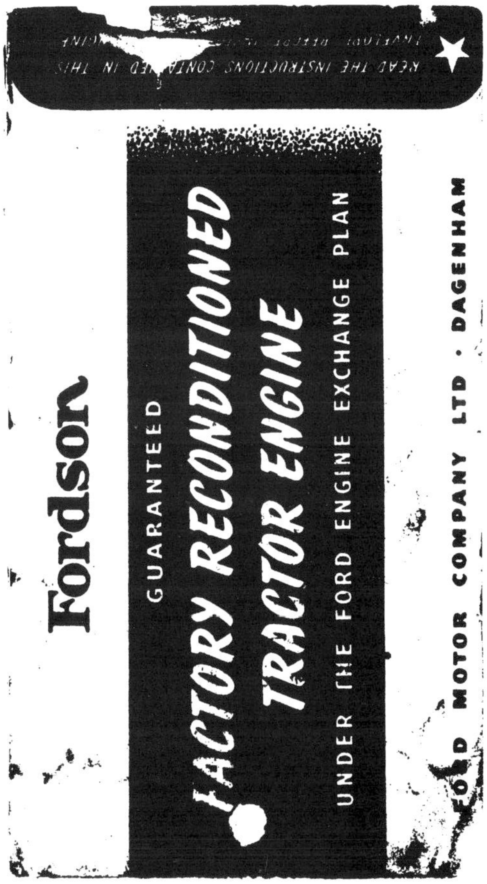

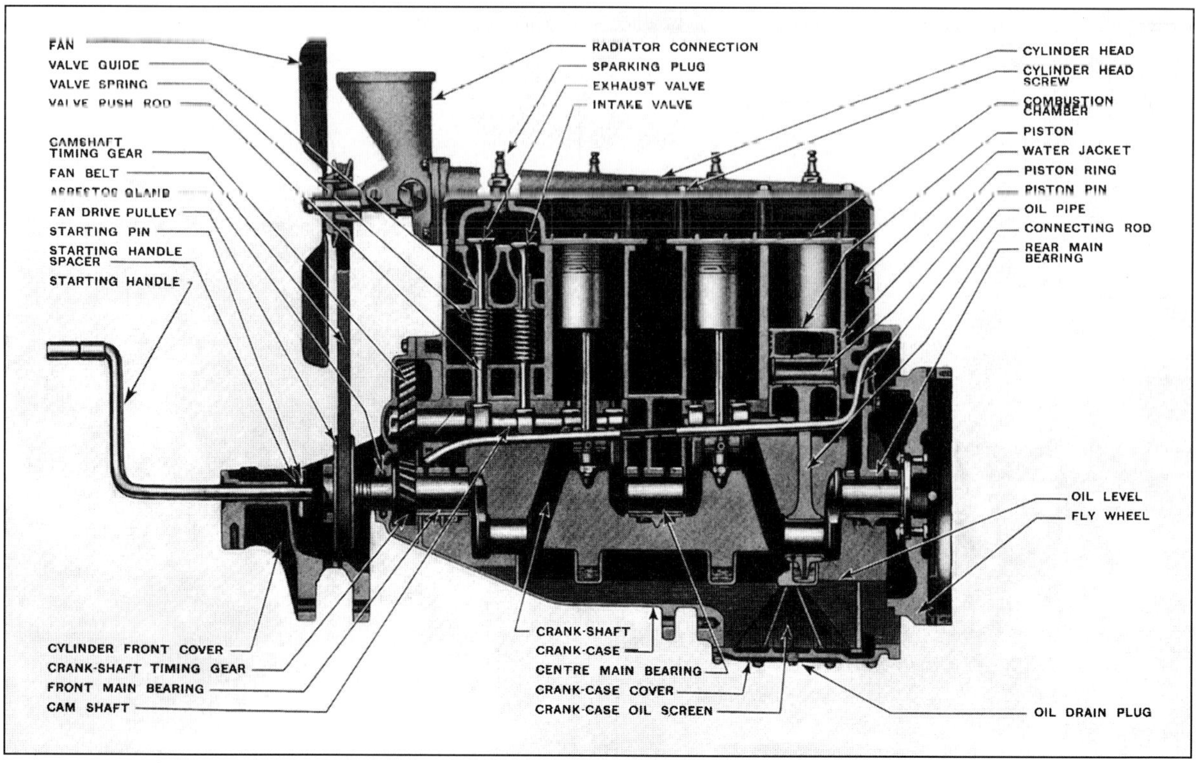

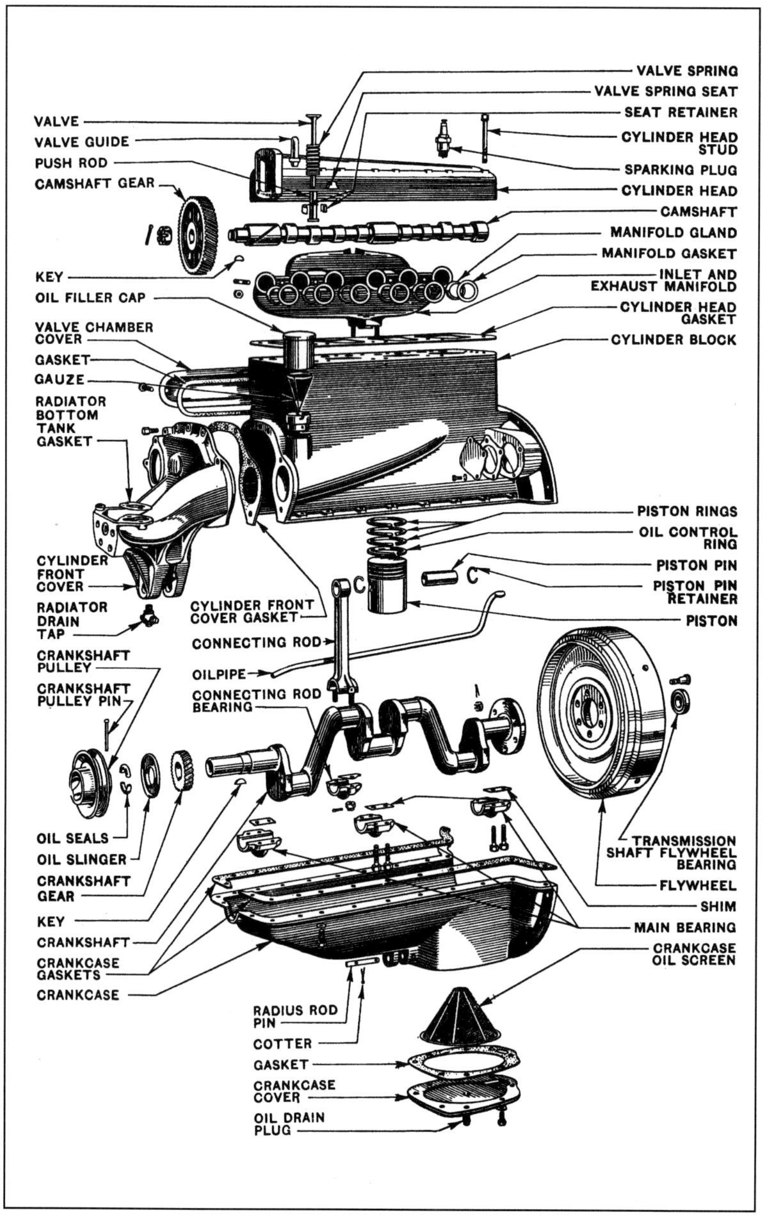

***Above:* The engine in section. It was a simple, basic design which did its job; it lacked sophistication to enable the easy overhaul methods which had come in on many other contemporary engines in the nineteen thirties, to be used.**

***Left:* The engine in section.**

***Opposite page. Upper left:* Diagrams showing the revised oil pipe used from tractor 1011911.**

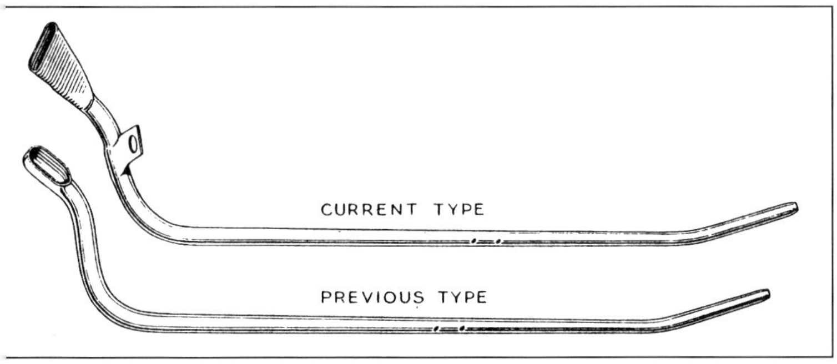

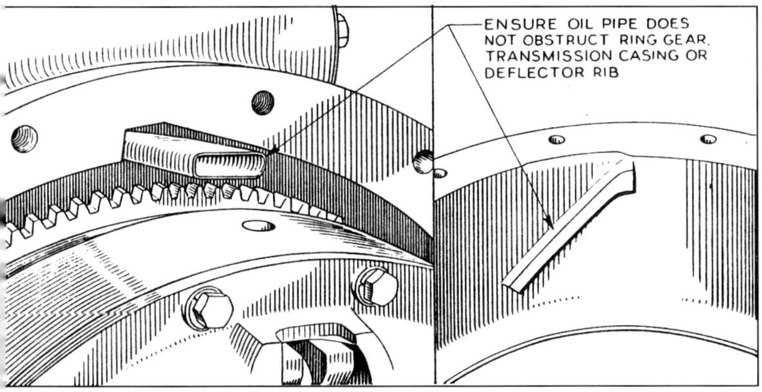

Top right: The single plate clutch used in the E27N with its pressure plate and slotted disc. Early clutches had plain discs.

Fuel Tanks.

The tanks used on latter day N models were taken over without much change, the main difference being the deletion of the brass instruction plate and holder and its replacement by a transfer. Tanks for Petrol only tractors had the filler at the rear. The rear filler on tractors for TVO or gas oil operation had the letters "Petrol for starting only" embossed into the RH side of the tank. The introduction of the diesel tractor caused the tanks for this model to have the filler located centrally - this was because the position of air and exhaust stacks made filling at the rear difficult. Late tanks from 1950 had an extra horizontal rib pressed into the ends.

Australian tractors had 'Major' transfers on each side of the tank.

Sediment Bulb and Pipe Line.

The old 'N' style brass sediment bulbs were used in 1945 only, being superseded by an AC pattern with glass sediment bowl and separate taps for Petrol and TVO. Original taps had gland nuts, but from 1950 rubber 'O' rings were used. The equipment for petrol tractors was simply the TVO version with the auxiliary Petrol inlet and tap position undrilled. An internal fuel filter was also added inside the tank from late 1949.

There are in circulation some glass sediment bowls with a single lever type changeover. There is no evidence that these were ever fitted in production and seem to be a proprietory replacement sold to dealers through the motor accessory trade.

Manifolds.

The hotplate type Vaporiser as used on the model N saw use on the E27N with little modification, the die cast mixing chamber with vacuum economy device having been introduced on that model in 1944. A manifold shield was always standard on TVO E27N's; the early tractors had no detachable slide, and the governor rod passed through the shield. With the moving of the control rod outwards on later 1945/6 tractors, the shield ceased to have the holes in it.

From 1947 an alloy float chamber became usual. For operation on Gas Oil the manifold was cut away and a bronze vaporiser plate fitted which had two halves, inner and outer. The width of this precluded the use of manifold shield until one was provided from 1949 on.

Petrol tractors inherited the same manifold as the 'N' using the downdraught Zenith carburettor. This carburettor was adapted for TVO use on the introduction of the New Type Vaporiser in production from 15th May, 1950. Many thousands of conversion kits were supplied to fit this equipment to both 'N' and 'E27N' tractors of earlier vintage; whilst starting and changeover problems were in theory eliminated, it could be said that the tractors never developed the same power as with the old style manifold properly maintained and adjusted, and sump dilution problems were rife. Early examples had a screw locking device for the heat adjustment lever; later on, in 1951 and on service kits sold after that date a latch was provided to hold the lever in any one of the four positions. The first tractor fitted in production with the 1950 Vaporiser was 1153615. Drain valuve deflectors were fitted underneath the manifold to prevent excess fuel from the manifold drain valves spilling on the exhaust. In Australia these were much longer and essential due to the high ambient temperatures experienced in the bush.

A large number of proprietory Vaporisers were available to fit the E27N. Loddon, Beccles, Elliott and Fishleigh & Hadden are to name but a few.

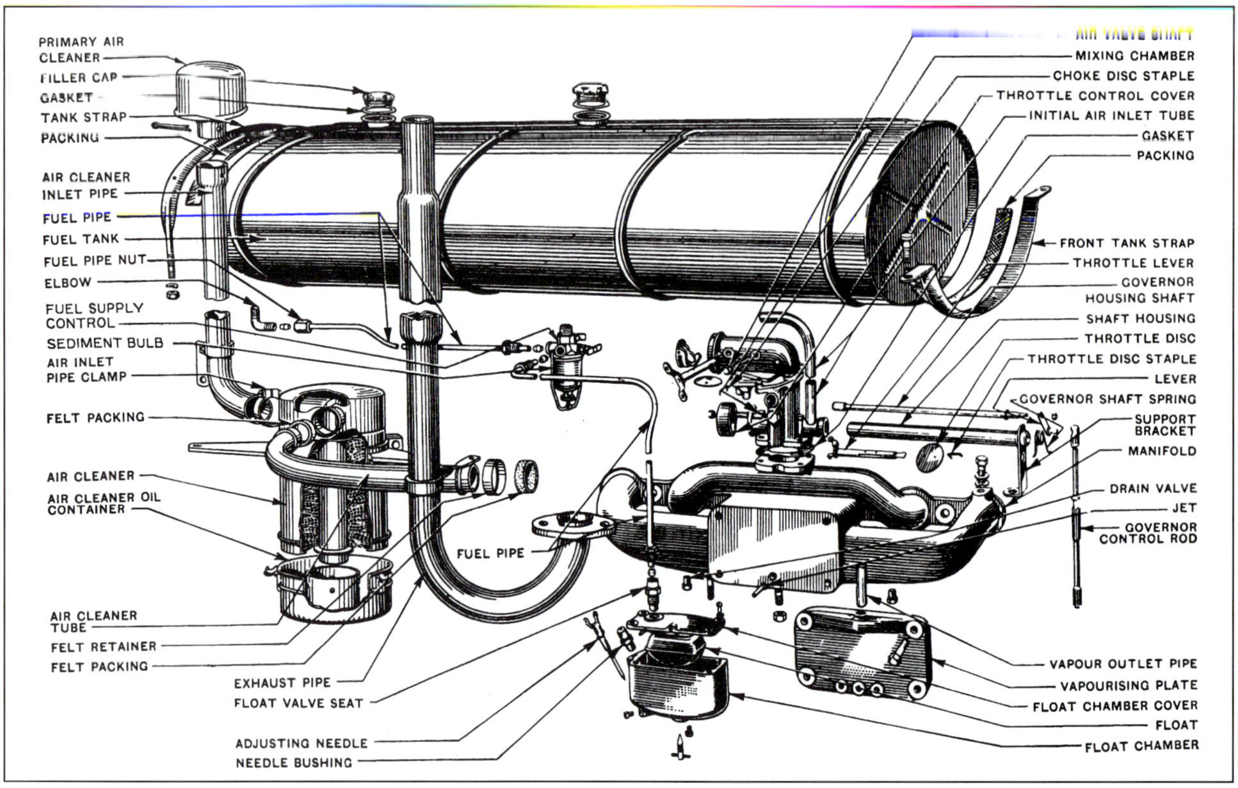

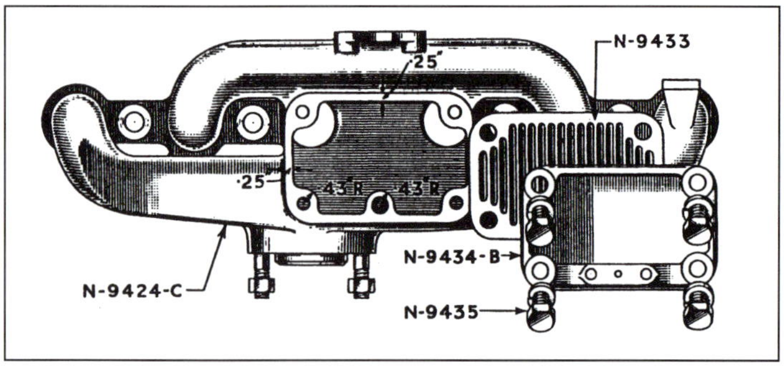

Above: ***An exploded view of the original vaporiser assembly.***

Left: ***For export, a Gas Oil vaporiser was available with bronze vaporiser plate. It was used with engines equipped with domed pistons and low compression cylinder heads.***

Below: ***How the Vaporiser works. It is shown connected to the Burgess type of oilbath cleaner.***

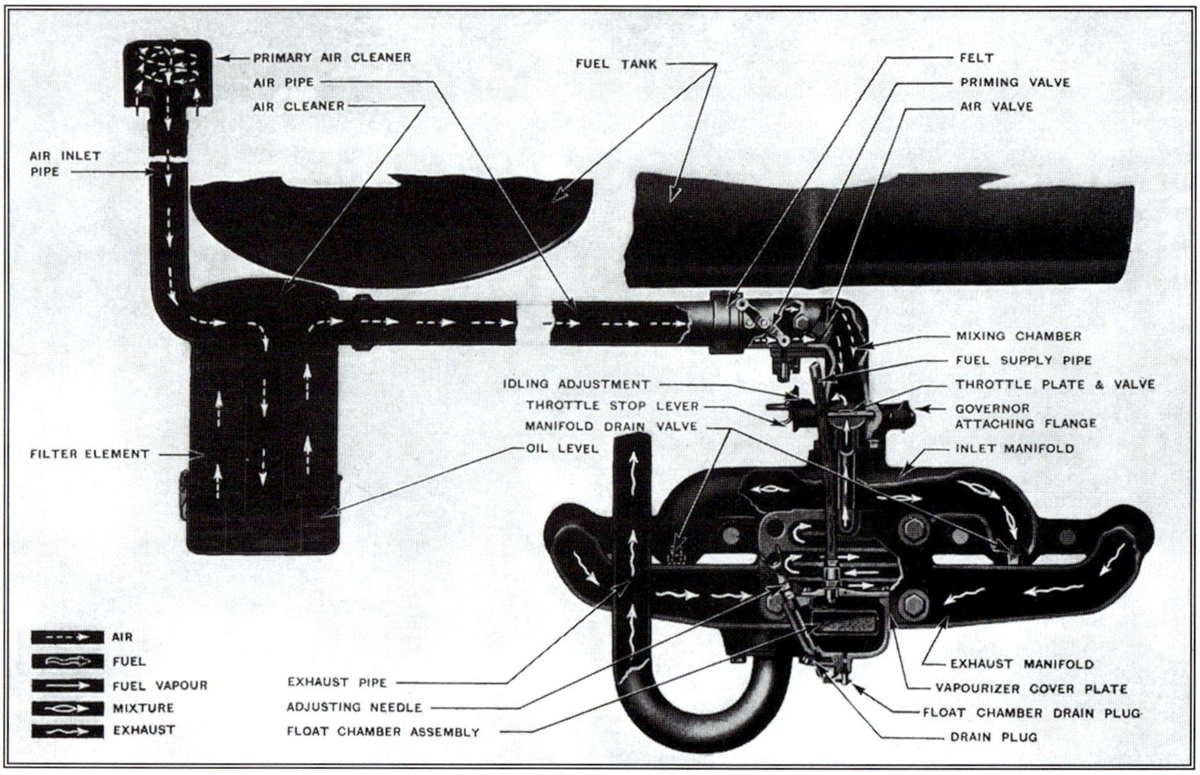

The Improved Vaporiser was introduced in production from 15th May 1950. It eliminated the maintenance problems associated with the hotplate type, and used the downdraught Zenith carburettor previously fitted to Petrol tractors, with the addition of a drain valve. Fuel lines, ignition conduit, governor linkage, air pipe, and exhaust all had to be modified to suit this new equipment. Whilst tractors so fitted always sounded 'sweeter' there was a definite loss of power with the new equipment viz a viz the old, and sump dilution problems increased dramatically - that is why the four bladed fan was reintroduced in 1951.

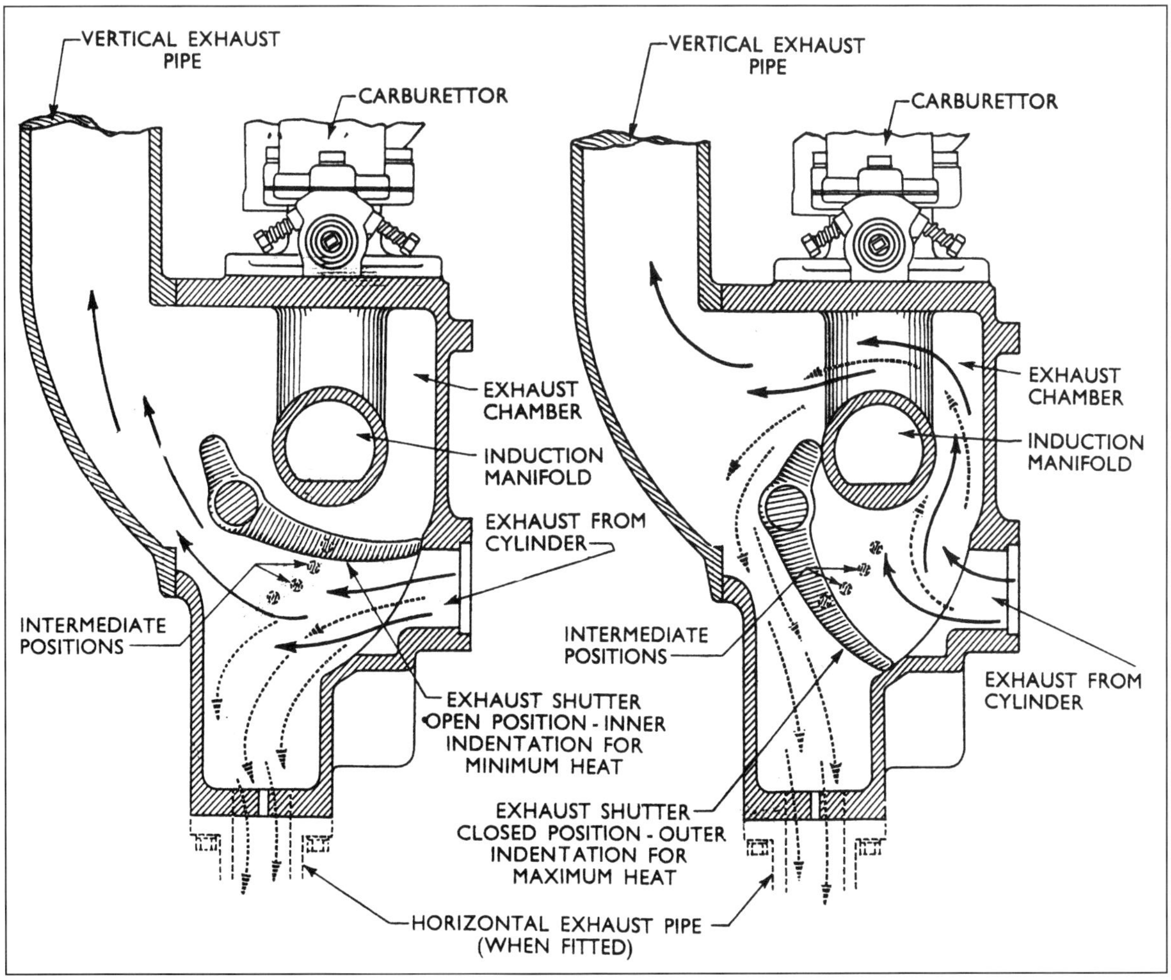

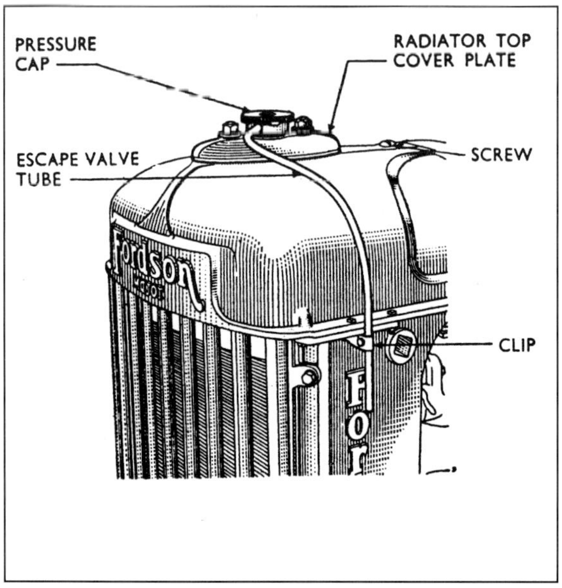

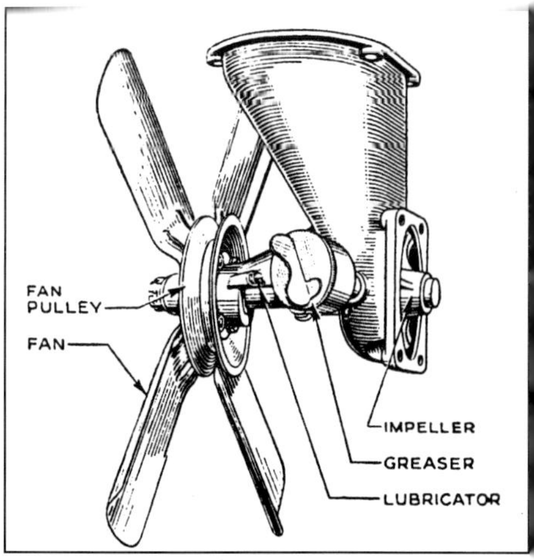

Left:* *The pressure cap installation.

Above:* *The water pump with four bladed fan fitted.

Below:* *The six bladed fan and radiator shroud used from 1948-50.

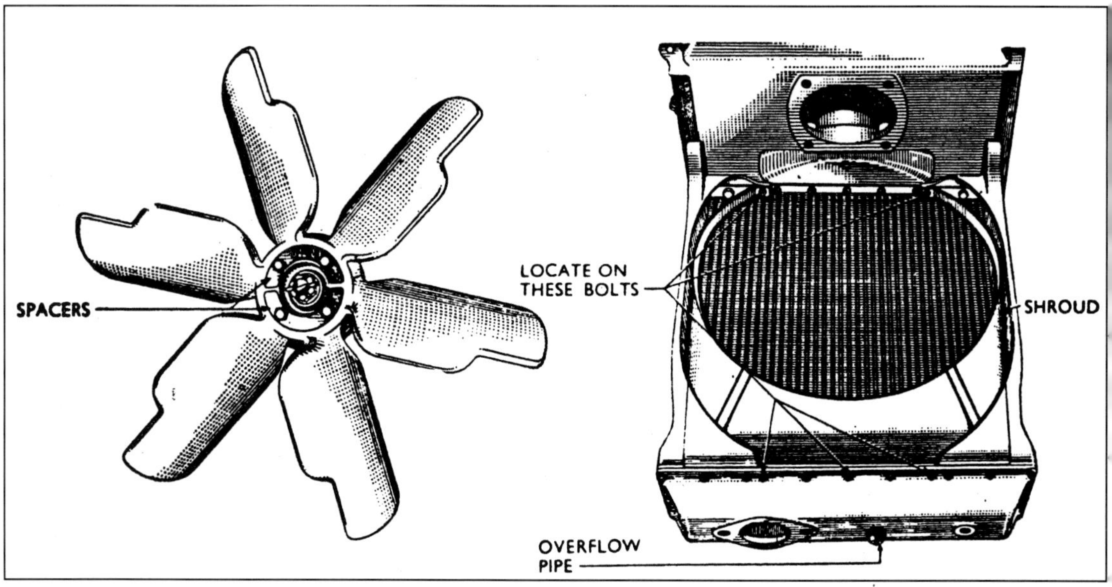

Cooling System

The initial prototype featured an 'N' style radiator, but soon the appearance of the tractor was altered by providing a different style of top tank and a grille to protect the core. The usual radiator blind fitted between the grille and the core. The very early top tank castings on the prototypes still had the filler cap athwartships, a lá model N, but this was quickly superseded by a design with the filler cap turned through 90°. The first 100 or so production tractors had this type of top tank, ther final design with swept back ribs on the side appearing after this. Early radiator side panels showed where the patterns had been altered from 'N' stock as they retained a little rib round the headlamp bracket holes.

It took some time before the cast alloy grilles were perfected or indeed in abundant supply, so some early tractors had fabricated ones. The design of the grille was altered soon after the introduction of the diesel engined tractor in 1948 to narrow bars, to give more cooling circulation, and this design was subsequently fitted to all tractors produced from that time.

The water pump design was, of course, inherited from the 'N' and was little altered. The introduction of a six bladed fan with associated shroud on the radiator in late 1948 was intended to reduce the possibility of overheating; however this equipment with its characteristic 'hum' at high engine speeds was sometimes over efficient, and the four bladed fan was reintroduced in mid 1951. This had something to do with the quick changeover characteristics of the improved vaporiser and the fact that sump dilution increased dramatically on tractors so fitted.

For export territories, a pressure cap installation was introduced in 1948. On this the overflow pipe was external, and the internal pipe deleted. Some tractors in service today may have a much narrower radiator core with an external overflow pipe to the back of the core; this is not of Ford supply, but Covrad made these cores available from the mid 1950s.

The arrangements for tractors fitted with P6(TA) engines required a different radiator bottom tank, and the offside radiator side member was cut away at the bottom to clear the dynamo pulley.

In the mid fifties, only the radiator core without overflow pipe was serviced by Ford for a period, and this required the addition of an external overflow pipe, an untidy and unsightly feature. The alternative was to fit the pressure cap installation. Protests from dealers and users soon found the core with overflow pipe inside reintroduced in service.

Australian E27Ns had radiator shutters fitted, along with a pressure cap installation. The normal calormeter could not be used with the pressure cap and was the location hole was normally blanked off, but in Australia a special type of temperature gauge was fitted.

A word on calormeters. There was a shortage of the usual Willmot Breeden type in 1945, and an AC pillar type was fitted on some tractors. These were by no means universal, and it would appear that 1945 tractors could have either. From 1946 however, the Willmot Breeden unit became universal.

Perkins Diesel engined tractors had the usual Perkins badge attached to the radiator grille. In Australia "37-41 Diesel" badges were also fitted.

CHOKE & THROTTLE CONTROL; GOVERNORS

The governor, mounted on the front RH side of the cylinder front cover, was inherited from the 'N' with only one change once the indirect throttle linkage was dispensed with, that is the Lever (Manual Control operating), on tractors built in 1945 where the governor control rod passed through the manifold shield, then through the upper dash and was bent out to meet the bracket on the inside, the lever was straight, on tractors from late 1945 the lever was bent and the governor control rod then passed on the outside of everything, being supported only by a bracket from the rear manifold stud.

The Choke control rod on early tractors was bent to clear the aircleaner; on later tractors it was straight.

Industrial tractors were available with a direct throttle. The lever was sited underneath the steering column and the linkage operated via ball ended joints and rodding to the front of the carburettor. Strangely the Improved Vaporiser was available with throttle control.

Where governors were not fitted the aperture was blanked off with a cone shaped plate.

AIR CLEANERS

The early E27Ns inherited the cast iron 'square' type aircleaner of the Fordson, probably to use stocks up. This item was expensive to produce, so Ford went out and bought in the standard 'Burgess' cylindrical oilbath cleaner made to suit the application. There appear to be two types of this in circulation, the earlier ones are not as deep and have less swagings around the circumference. The primary air cleaner was of the customary 'saucepan' type as inherited from the Fordson. Early diesels had a similar type but with four flat support vanes. This was superseded by the large domed pattern by the time diesel production had really got under way, by December 1948 in fact.

Sporadic use was made of a cylindrical cleaner of AC Sphinx design, mainly in the 1948-52 period. This may have had something to do with a fire at Burgess' works in Hinckley. The Sphinx cleaner was of lesser diameter than the Burgess and had a flat top, whereas the Burgess had a slightly domed top. The bracket from the air pipe to bash was originally secured by only one bolt, but the habit of climbing on the tractor by standing on the drag link and hauling oneself up on the aircleaner pipe often caused this bracket to fail; late tractors from 1949 had an improved bracket which attached to the dash with two bolts.

The pipe from Aircleaner to Carburettor was of steel, with felt packed retainers, until the introduction of the 1950 Vaporiser, when a rubber hose was used. A special steel pipe was available in service to suit 1945 tractors with the square aircleaner fitted with the 1950 vaporiser. Industrial tractors used the Vokes dry type filter attached directly to the carburettor intake. Some export tractors, especially those sent to Canada, were fitted with Donaldson 'Ball Special' Glass Jar type primary air cleaners.

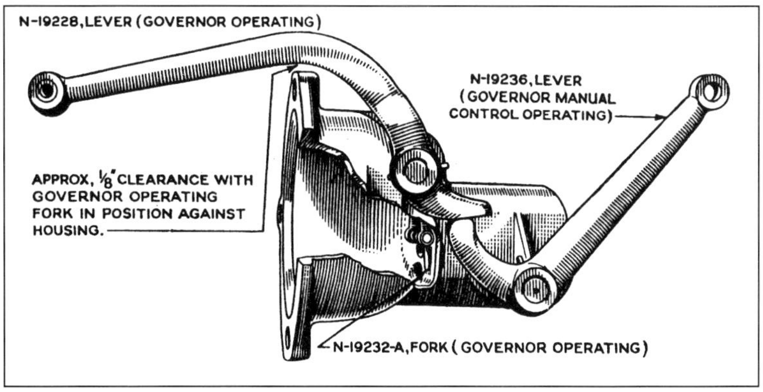

Right: ***The original pattern of governor housing with straight lever to suit the offset throttle.***

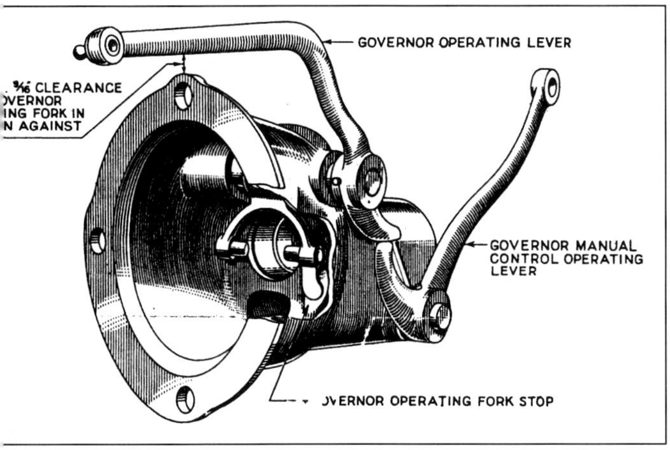

Left: ***The later type housing (these were cast alloy) with bent lever to suit the direct throttle rod. Where the improved vaporiser was fitted the ball on the governor operating lever was refitted to the outside.***

Exhaust Systems.

The original exhaust system of the E27N was inherited from the Fordson and lasted only up to serial number 983125. This only accounted for the first six weeks production, during which time Burgess of Hinckley were tooling up to produce thousands of that familiar 'sausage' silencer which graced the remainder of E27Ns with vertical exhausts. The Industrial tractor used a modified version of the tapered horizontal silencer first used on the model 'A' car and later modified for use with the 'N' tractor, and this was also available as an option for all other models. The introduction of the Improved vaporiser in 1950 saw the vertical exhaust become a straight pipe 34 1/2 inches long which dropped into the heater box cover. If you wanted a horizontal exhaust with this fitting the bottom of the manifold had to be drilled and tapped, and a horizontal silencer of the usual tapered variety fitted along with a flat cover plate. Later Australian tractors had Spark arrestor exhausts fitted.

ELECTRICAL SYSTEM

Some confusion is evident as to the first electrical equipment fitted to E27N tractors. It would appear that until early 1946, if lights were required, the model 'N' six volt system of lighting only with the usual three brush generator and the battery slung on the side of the cylinder block, and controls in the same position as the model 'N', i.e. the switch under the radiator operated by a rod back to the dash. The early parts list does not show any of these parts, however. Eyewitness accounts at the time differ as to whether straight or bent headlamp brackets were used.

By early 1946, availability of materials and components meant that the 12 volt electrical system designed for the tractor could be introduced. From 1st March 1946 all tractors were equipped with a starter ring gear whether electrics were fitted or not.

A two brush generator, 4.5" in diameter was fitted and driven from the fan pulley. From March 1949 a 3.9" generator was fitted to spark ignition tractors but the 4.5" diameter type continued in use on Perkins Diesel engined machines. The voltage regulator was originally housed in a metal case on the outside of the dash. The battery was housed on a platform attached by three bolts to the upper dash. The instruments were housed in a semi-circular box attached to the steering column and consisted of starter and horn buttons, lighting switch and ammeter. The starter motor was of the indirect type with solenoid attached. Headlamps were E93A pattern with 12 volt bulbs fitted, giving main beam, dipped, and side light illumination, and the headlamp brackets were cranked to bring the lights to a suitable height. Early headlamp brackets were lighter and not as long as later ones - no exact date has been found for the change. The horn was mounted on the RH headlamp bracket; it varied in design depending on the supplier. A tail light and socket were fitted just under the seat to the top of the rear transmission housing.

From March 1949 a direct type of starter motor was fitted and the instrument panel was redesigned as an oblong box, now housing the direct starter button, the voltage control regulator, as well as the other items. It was possible to convert to the new type of starter and still use the old instrument panel. The service letter for December 1949 shows the direct starter switch fitted to an earlier tractor on a fabricated bracket attached to the lh side of the upper dash.

Diesel tractors had different arrangements, with two six volt batteries slung either side of the steering column and the control box mounted between them longways.

The conduit which carried the wiring to the front of the tractor was initially held in position by three clips brazed to the tank. From 1949 however, a new style which clipped under the tank at the rear and was held up by the tank strap at the front was introduced. A second conduit attached to the clutch housing carried the main feed to the starter motor..PA

MAGNETOS

There were only two types of magneto ever used in production on the E27N, and these were both made by Lucas. The GJ4 magneto was used until 1947; the 4RF marine type after this. In theory the new magnetos came in at tractor No. 1021061 but it would appear that some later tractors up to about 1050000 could have the earlier type. The Lucas SR4 magneto was never used during E27N production, it was a cheap and unsatisfactory replacement available from the mid fifties. The lack of manual advance and retard always caused tractor engines so fitted to 'knock'.

There were two types of ignition conduit, one with a straight end used until 15.5.50 and thereafter only with petrol & gas oil engined tractors. That to suit the 1950 vaporiser had a bend at the end as the whole assembly was mounted further forward to clear the larger inlet manifold.

MUDGUARDS

A look in any edition of the Fordson Major Instruction Book will reveal illustrations of tractors with a type of mudguard never used in production. These fenders, made of one piece of sheet steel swaged into shape and supported by C section were similar in construction to those used on the New Fordson Major from 1951 on. The cost of making press tools to form this item in production was prohibitive, and thus only the few prototypes had such attractive mudguards. A compromise was reached where only the top was rolled (using the same tools as the Fordson 'N' utility mudguards were formed from, and then spot-welded to the sides. The whole ensemble was supported by an upturned 'U' shaped bracket and the mudguards were strengthened by having two ribs swaged in the shape of a 'V' on the sides.

This type of mudguard lasted until late 1945, when such items were being subcontracted, and an improved design was developed where the sides were supported by steel angle. These were universal until late 1948, and were made across the road at Briggs Motor Bodies. When Sankeys at Wellington became involved in E27N mudguard supply, a third design appeared where the mudguard side was supported by two box section pressings spot-welded to the sides. Briggs soon followed with an almost identical design but the Sankey ones had straight tops to the supports whereas the Briggs ones had the lugs bent in at 45.°

The footplates were of similar design throughout, but with the introduction of the third type of mudguard in late 1948 the construction of the fender platforms was altered.

The industrial tractor had a body assembly built by Briggs Motor Bodies. The canvas covered cab was optional, the basic tractor being supplied only with windscreen. The front end of the body had to be modified to suit P6(TA) engined tractors. Industrial tractors were also equipped with a vacuum operated windscreen wiper and a speedometer.

TOOLBOXES

Early tractors had the small 'N' type toolbox fixed to the dash. With the advent of pneumatic tyred tractors the large toolbox was sited under the seat. This location was fine provided that no lift was fitted; with the introduction of HPL equipment the first Smith lifts had location for bolting the toolbox to the top. However, it soon became evident that the best place was to relocate the toolbox on the side of the engine block; which was just a return to the same location as the model 'N' Land Utility.

Tractors less HPL still had the toolbox under the seat until 1948, from which time all tractors had it located on the side of the engine. Diesel tractors had the box fitted to the inside top of the lh mudguard.

Tools provided included a Jack, Grease Gun, Foot operated tyre pump, and various spanners and the famous 'Ford' pliers. A spark plug type tyre inflator was provided from 1948 on s.i. tractors.

SEAT

A deeper seat pan than that used on the 'N' was adopted for the 'E27N', although a few early tractors escaped with 'N' type seat pans. The spring was of different depth when the HPL was fitted. Industrial tractors had a motorcycle type seat, but with back.

ACCESSORIES

A rear view mirror was fitted to all pneumatic tyred tractors to the RH upper dash by one bolt. LU. and pneumatic tyred RC. Tractors without electrics also had a Lucas No. 12 Bulb Horn bolted to the outside LH upper dash.

The side drive pulley was the same unit as fitted to the 'N'; the same choice of Compressed fabric or cast iron pulleys was available.

Up to 300lb of wheel weights could be attached to the rear wheel hubs on each side, and a dual rear wheel conversion set was available.

HYDRAULIC POWER LIFTS

Two types were produced for Ford, one by S. H. Smith and the other by Varley. The drive was taken off the top of the gearbox. The Smith lift was a single lever control, the Varley two lever, the second lever operating external services. The Varley unit was more compact than the Smith and required a longer PTO drive shaft.

The three point linkage was originally non adjustable with connection to the lift arms by chains. With the introduction of a drawbar adjustable for depth the next logical development was to make the links solid and adjustable, with a levelling box on the RH link. Early Smith lift castings were Cast Iron, later ones were Alloy, from 1948. The lower link brackets were fitted to all tractors from February 1948 whether they had HPL or not, and the PTO shifter lever had to be bent to suit tractors with HPL. This bent lever was fitted to all tractors from 10th June 1948. If you wished to fit HPL to tractors before February 1948 the lower link support brackets had to be welded on.

Many E27N components continued in use on the EIA series of New Fordson Major.

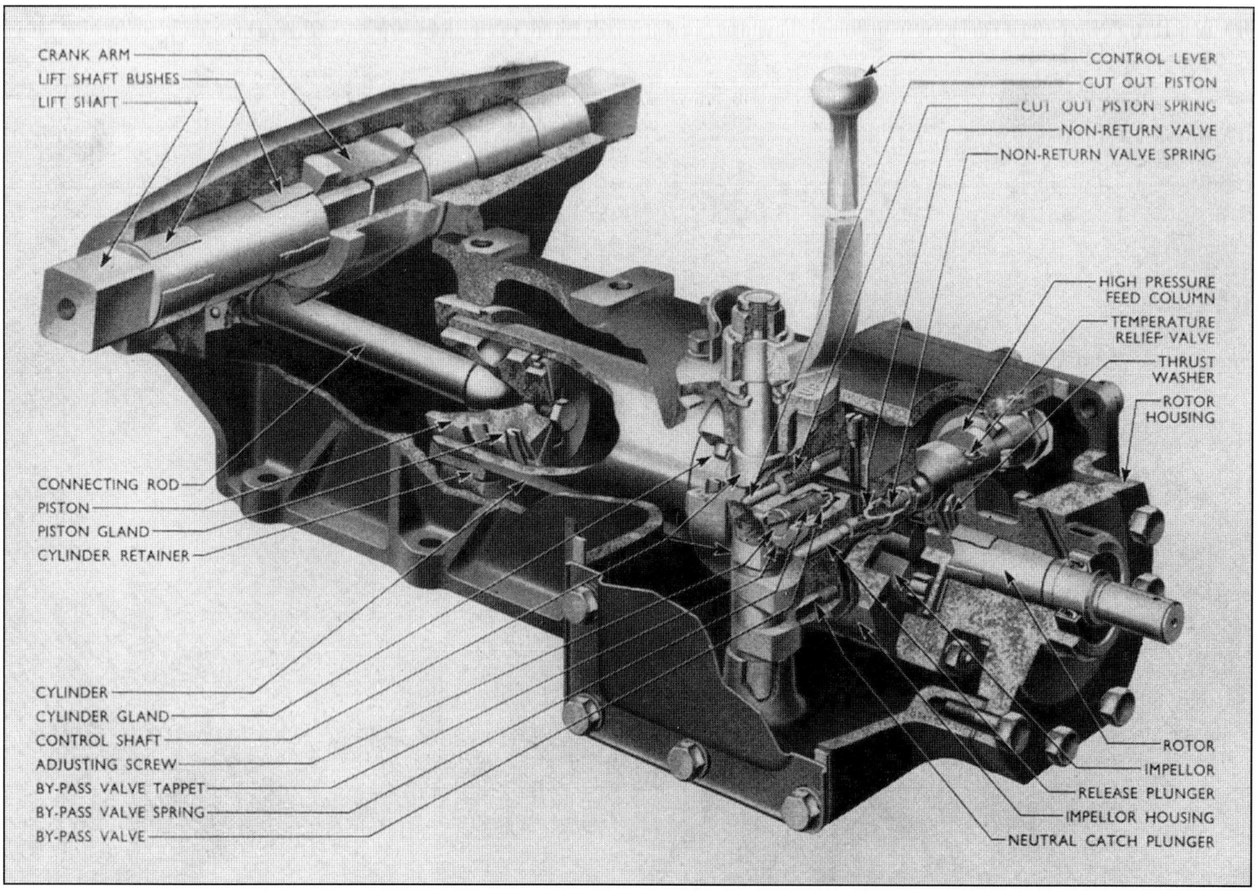

Above:* *The Smith single lever hydraulic power lift in section.

Below:* *Operation of the Smith lift explained.

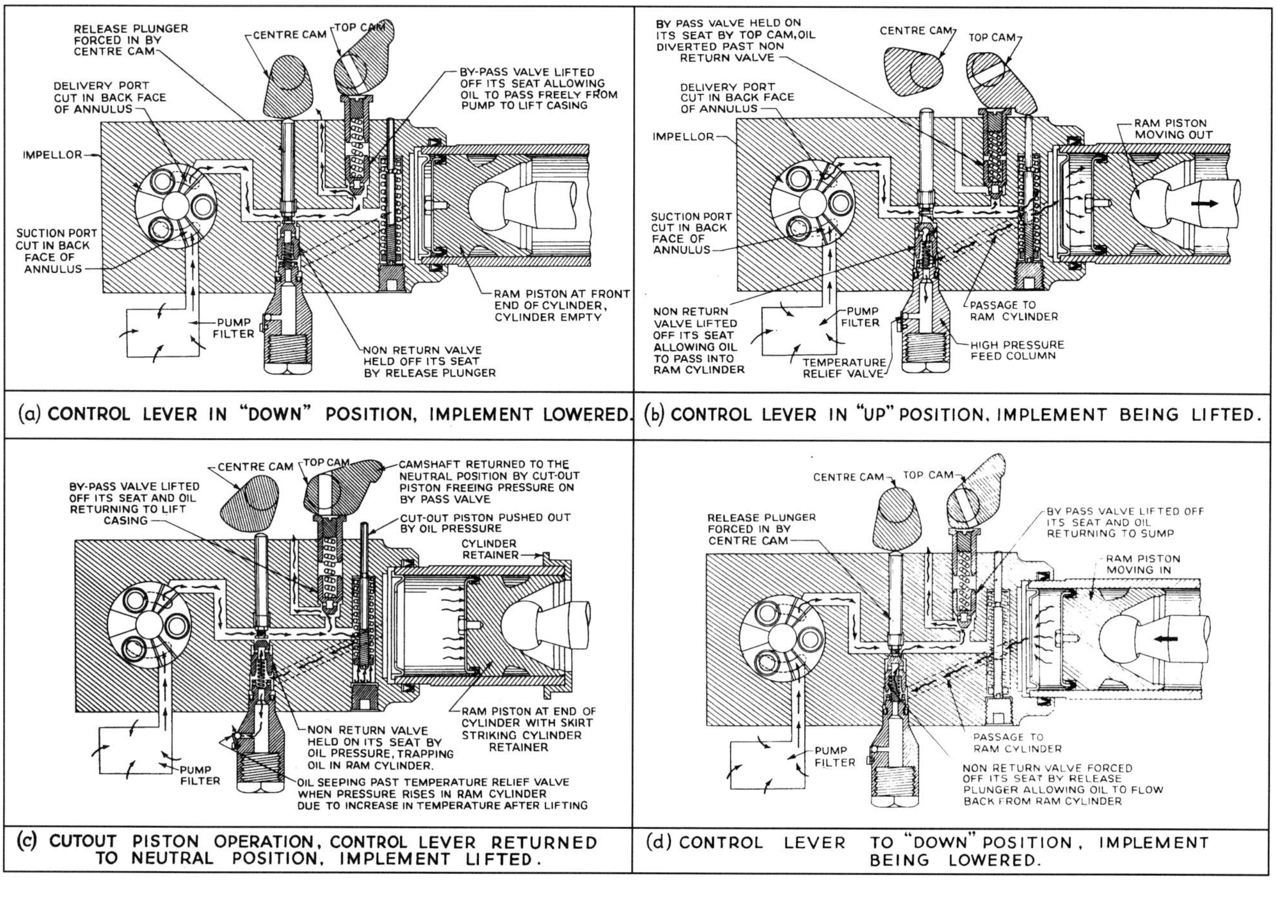

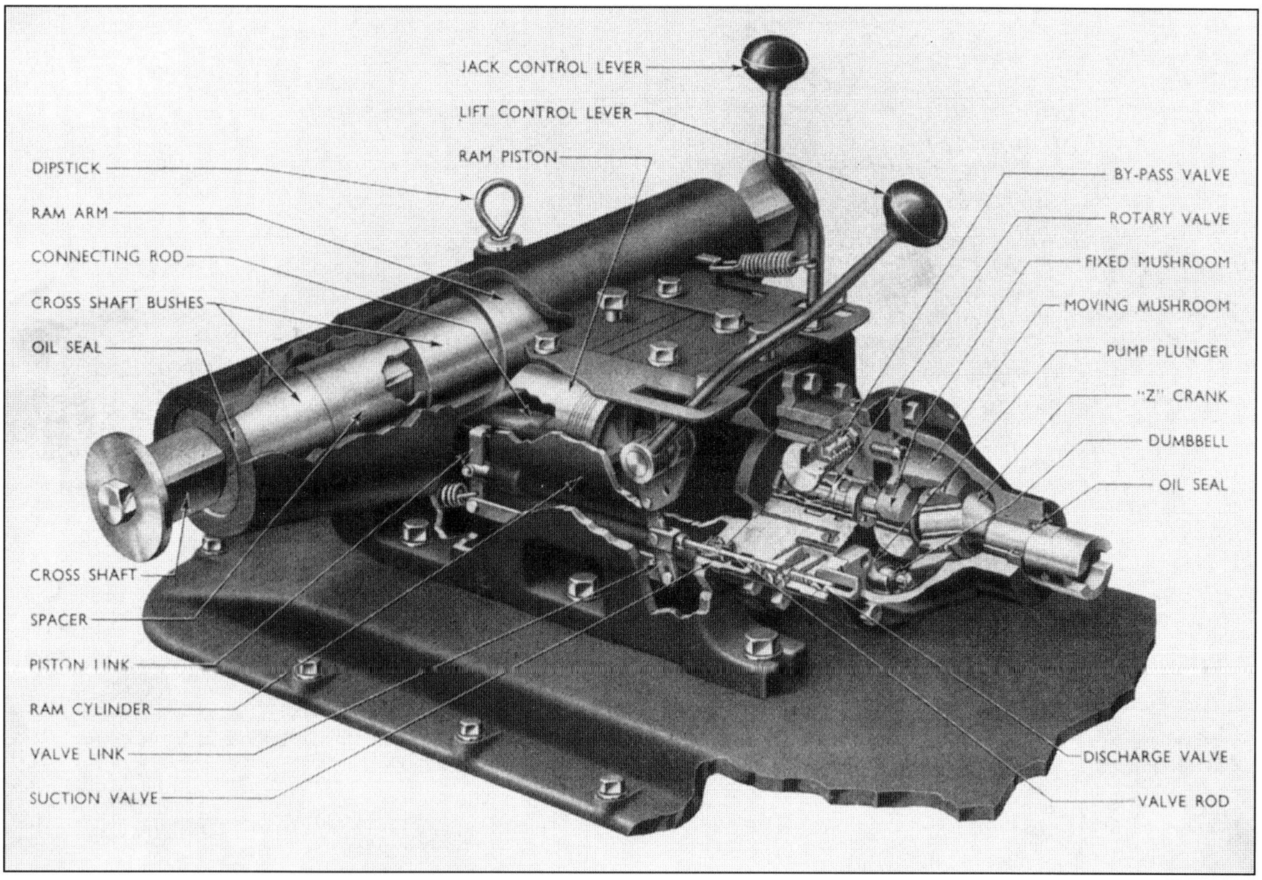

***Above:* The Varley two lever lift in section. On this type a separate lever was provided for the operation of external services.**

***Below:* Operation of the Varley lift explained.**

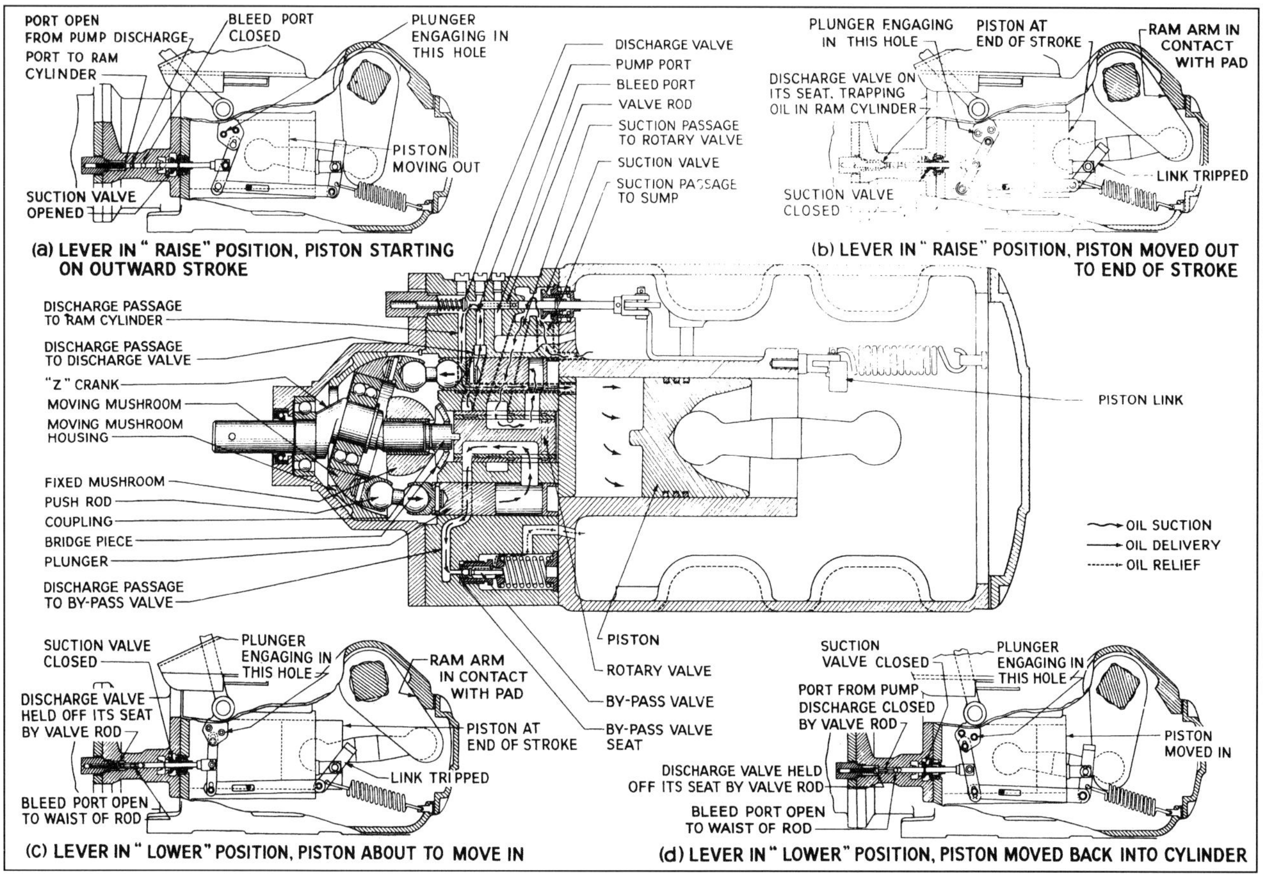

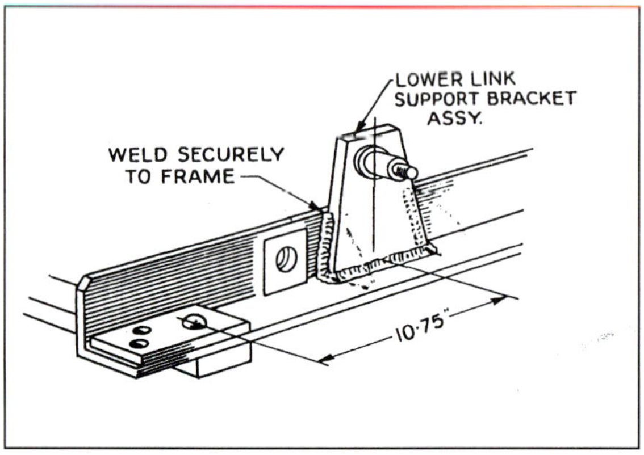

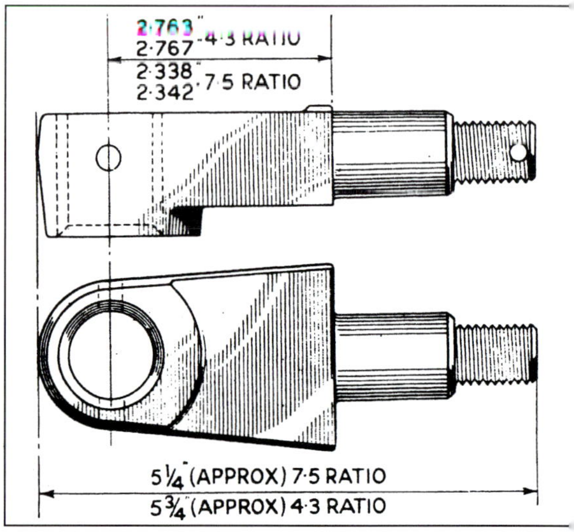

Above Left: ***The position of the lower link support brackets on the tractor drawbar.***

TRANSMISSION			LIFT DRIVE ASSEMBLY			
Gear	Identification Markings	No. of teeth on Transmission Low Gear	No. of teeth Idler Gear	No. of teeth Drive Gear	Length of Idler Gear Supports	Length of Drive Shaft
LOW TOP Standard Ratio .. LOW TOP Special Ratio ..	22 43 SR 43	21	26	29	5¾"	*13.84 or †18.44
HIGH TOP Standard Ratio .. HIGH TOP Special Ratio ..	22 75 SR 75	27	23	29	5¼"	*13.84 or †18.44

* Single lever type lift. † Two lever type lift.

Above Right: ***Different lengths of PTO idler gear supports were needed to suit the 4.3 and the 7.7 transmission.***

Below: ***The HPL drive assembly was based on the old 4.3 gearbox side PTO of the Fordson with an extra lug added and differing internal gear sizes.***

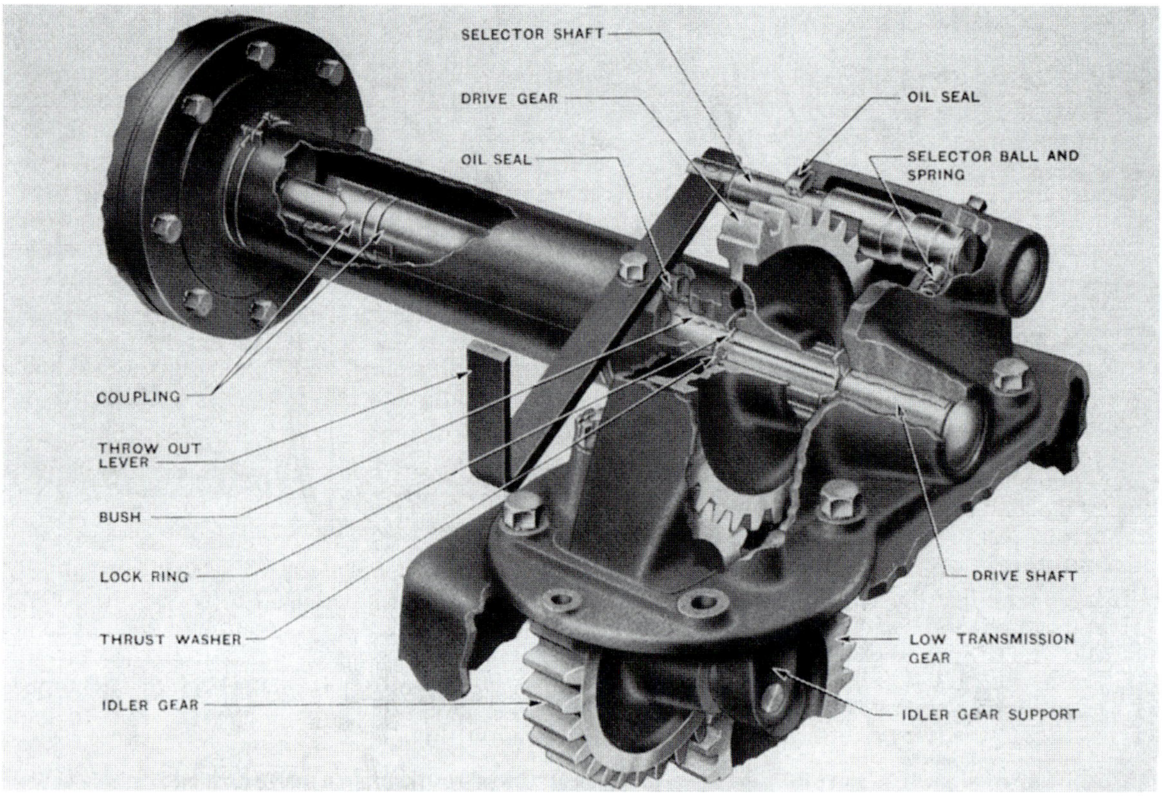

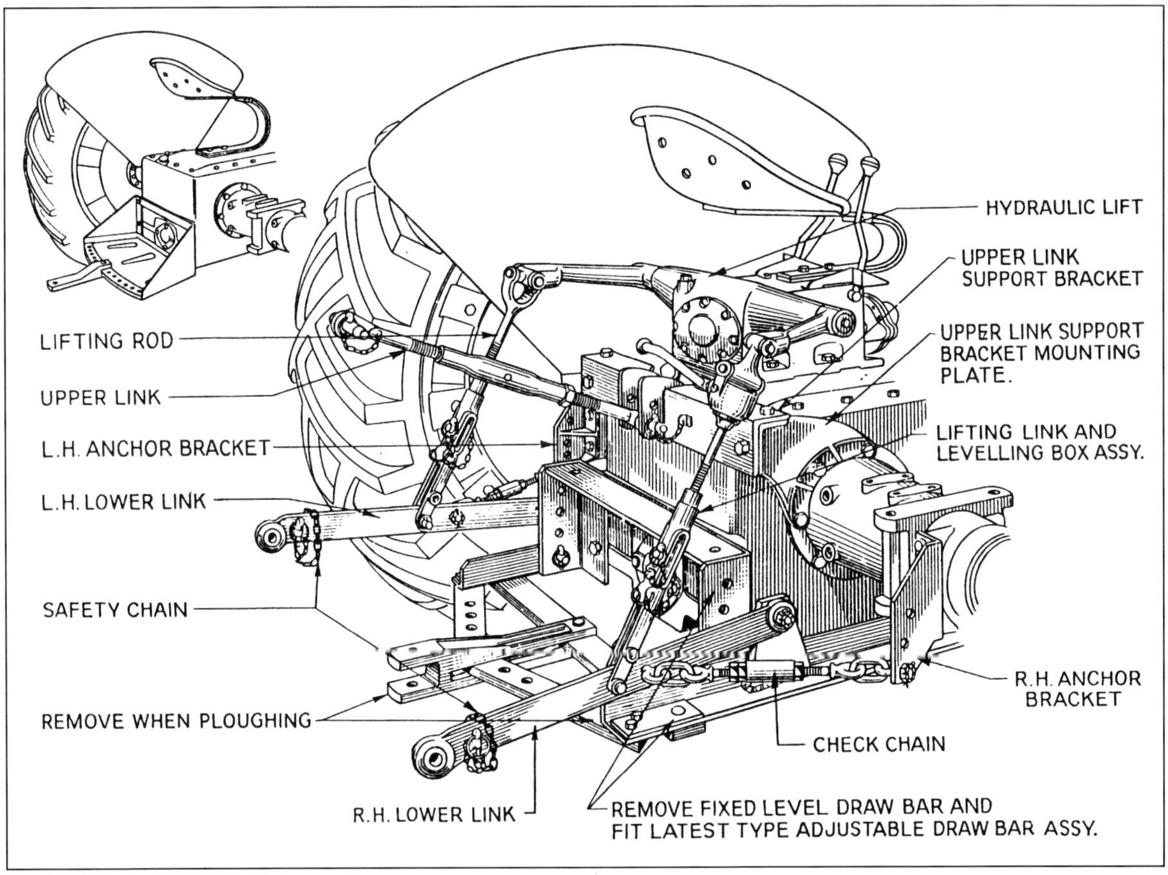

***Above:* The complexities of E27N lift linkage are such that we illustrate all the permutations here. Here a 1945 tractor has been fitted with an installation to allow ploughing more than 8" deep.**

***Below:* Early tractors with HPL had chain linkage (small illustration). If ploughing up to 8" deep it was possible to convert this to rod operation retaining the original fixed level drawbar.**

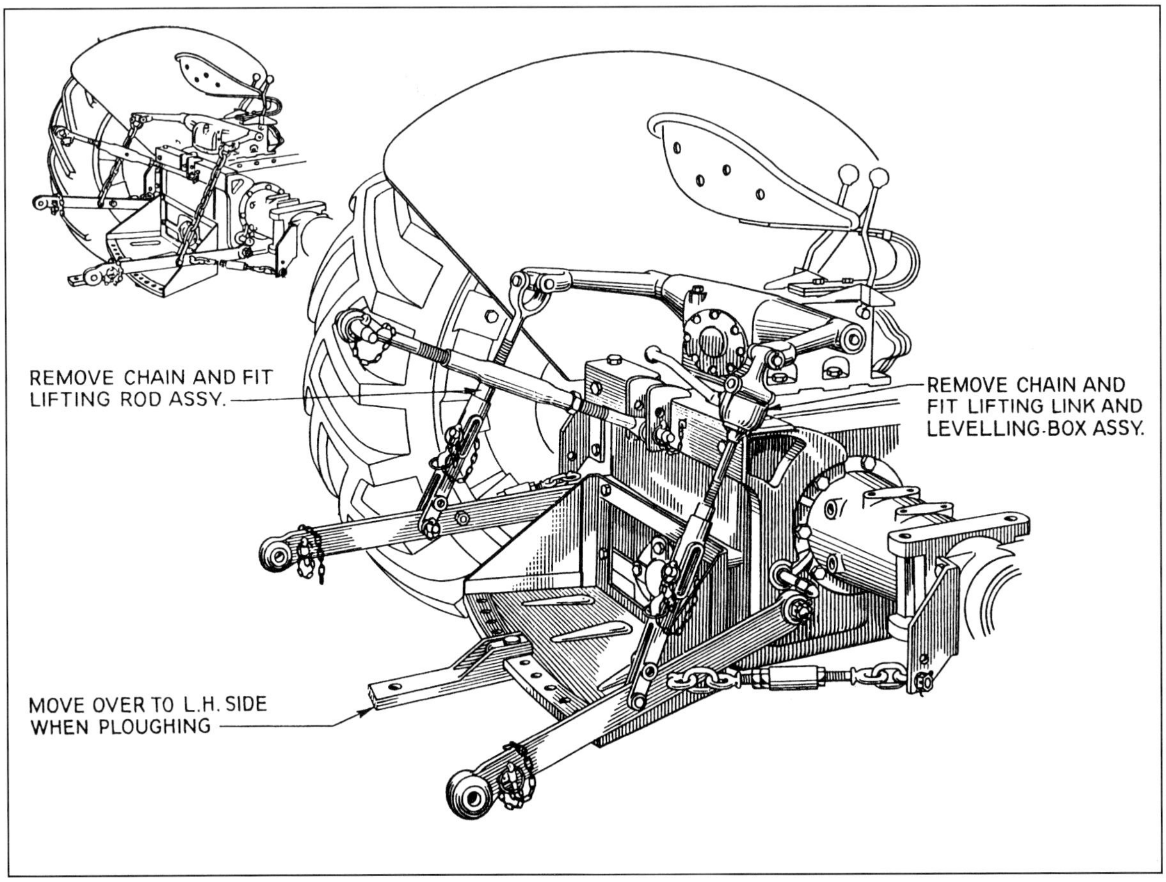

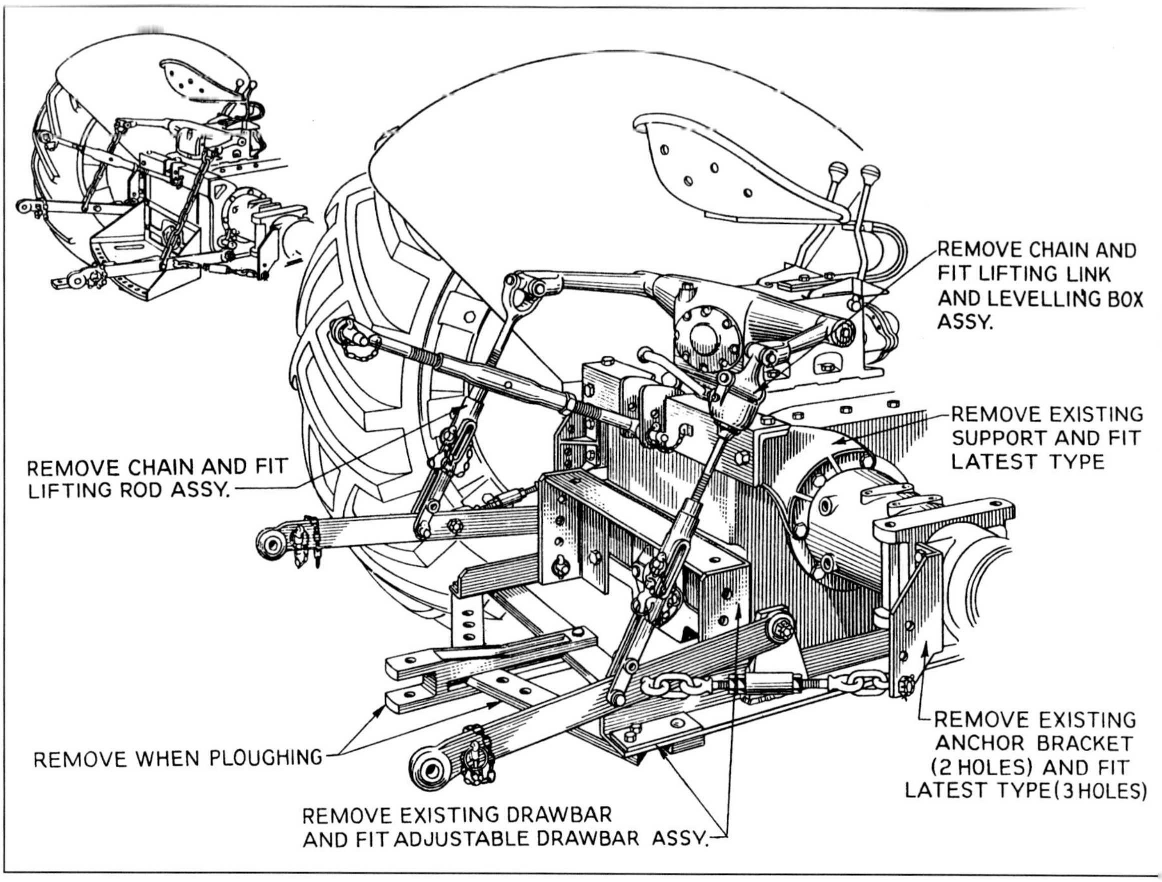

Above:* *If however, it was desired to plough more than 8" deep it was necessary to fit the adjustable drawbar and other new components as shown here.

Below:* *Where an adjustable drawbar was fitted and no previous linkage, it was necessary to weld the lower link support assemblies to the drawbar and fit the latest type of guide bracket with 4 holes.

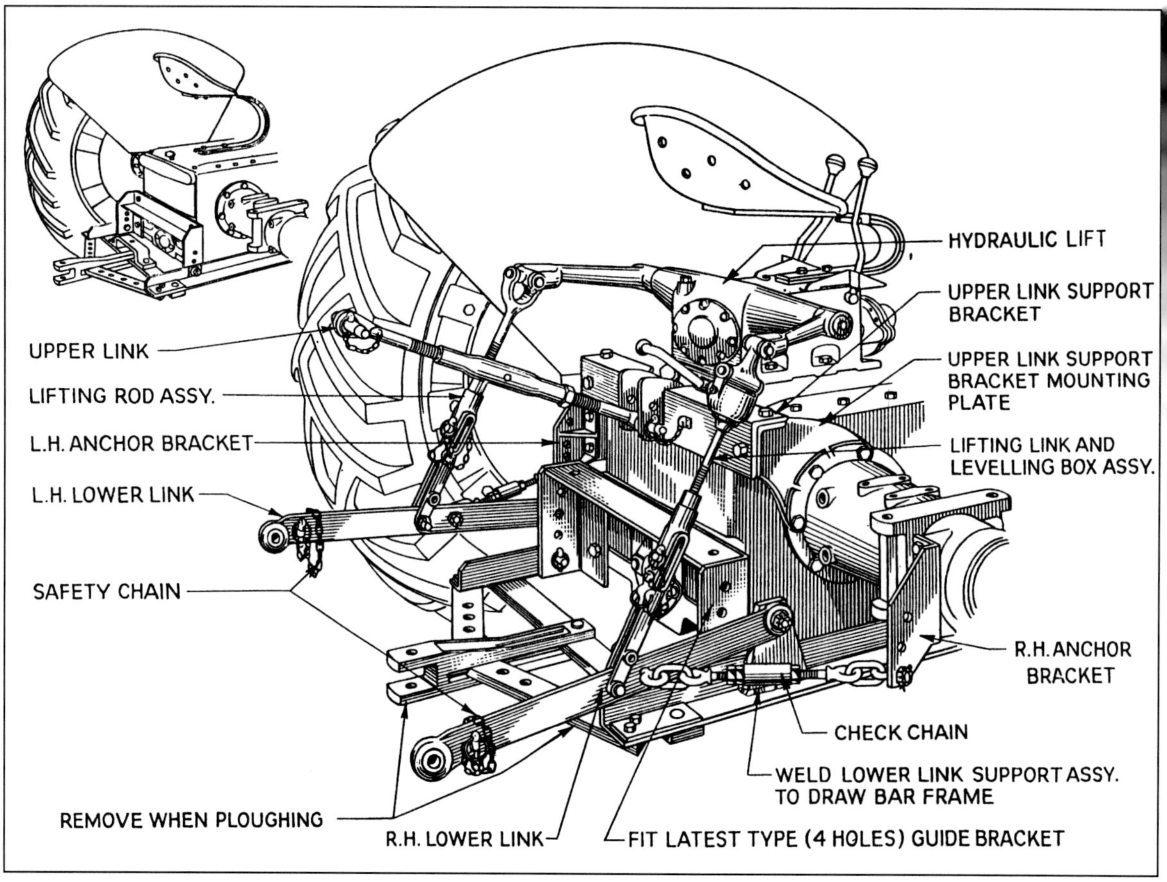

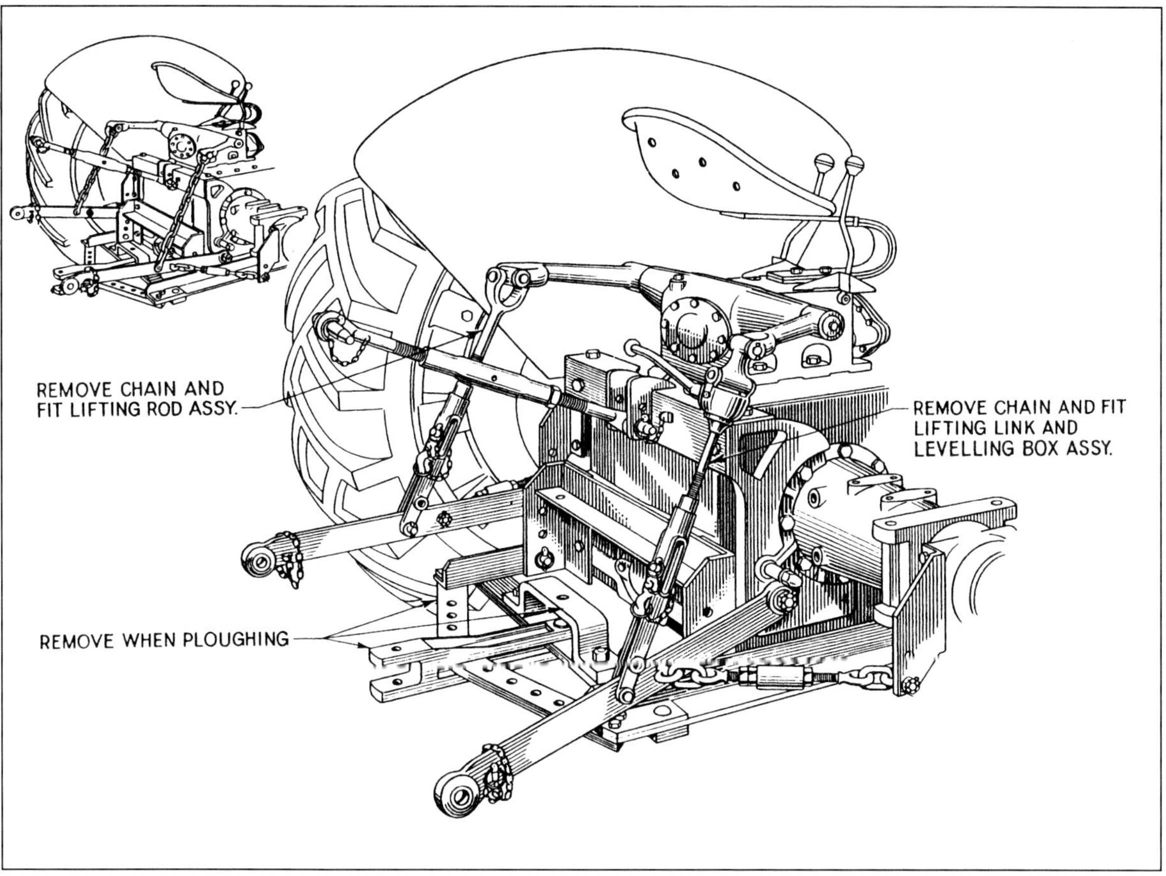

Above: ***Where chain linkage was fitted with the early type of adjustable drawbar if it was only necessary to plough up to 8" deep all that was needed to convert were the lifting links.***

Below: ***Where it was desired to plough more than 8" deep however, the HPL linkage had to be converted to the latest type.***

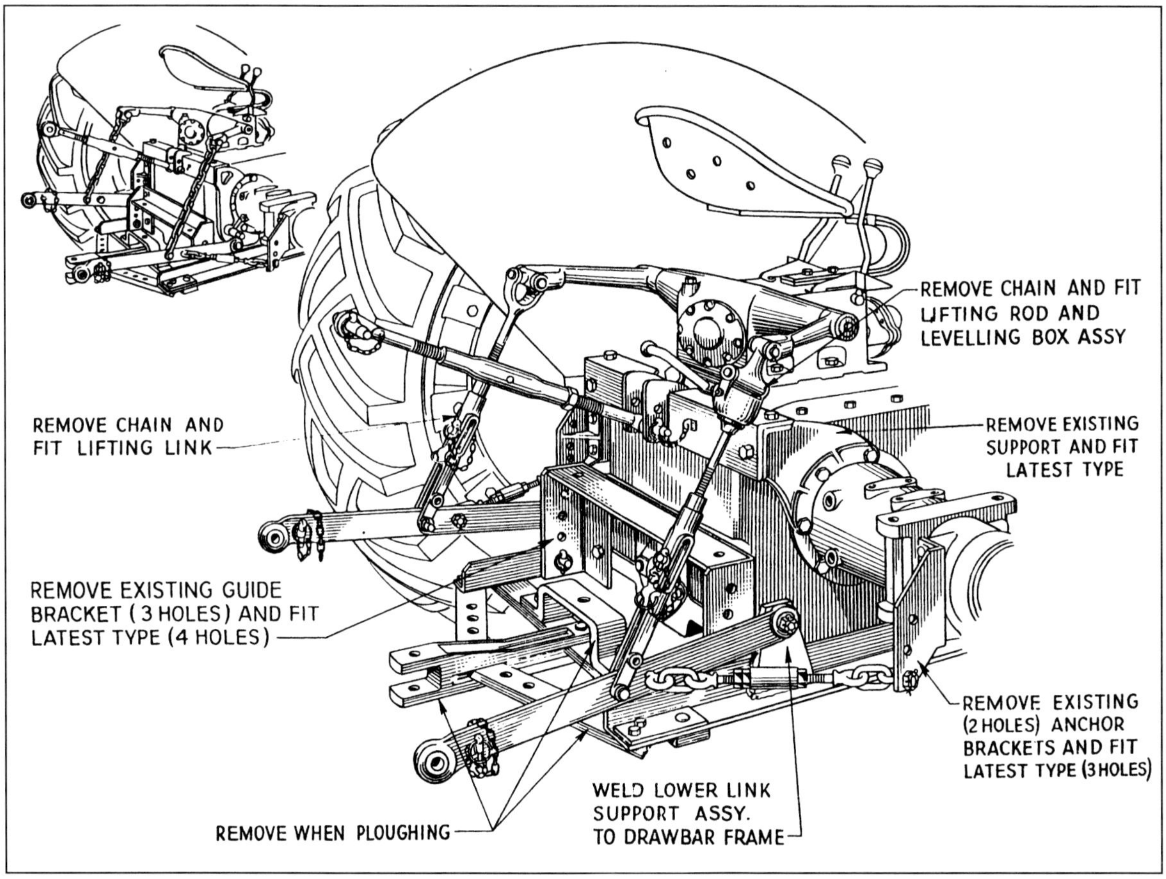

1945 PROTOTYPE

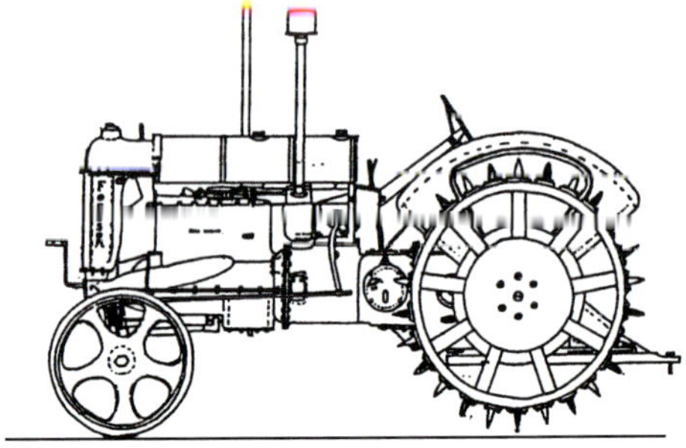

The first hand built prototypes used many stock 'N' parts. This included the fitting of an 'N' radiator, a feature which was copied by the artists when the produced the first cut away drawing of the tractor. Although the first prototype would appear to have been painted green, evidence now to hand shows that all other E27Ns were blue and orange.

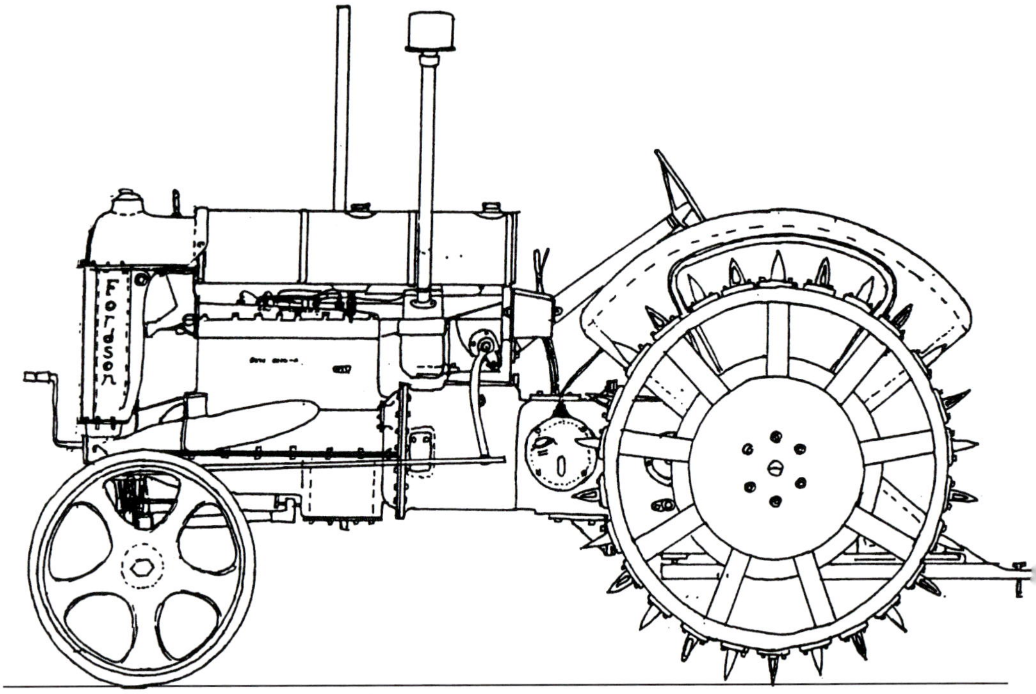

1945 PRE-PRODUCTION

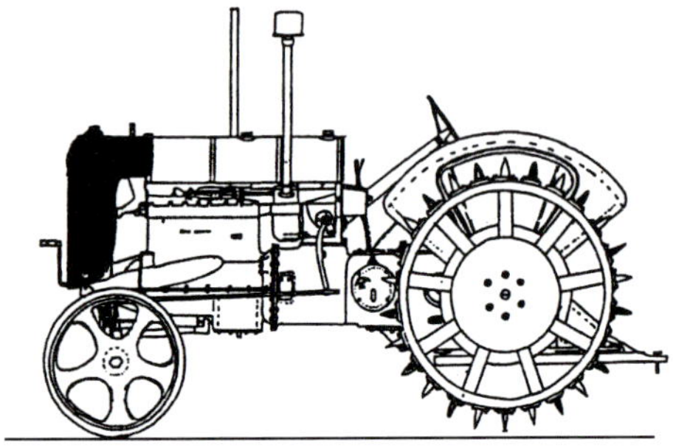

The desire of the sales people to 'modernise' the appearance of the tractor could only be really achieved by extensive use of tinwork, which was both expensive to tool and produce. By adding a front grille and restyling the radiator top tank it was intended to alter the appearance of the tractor sufficiently to make it look 'new' without any great expense. The first castings used, however, on pre production tractors, had the radiator filler cap cross-ways; the first grilles were fabricated.

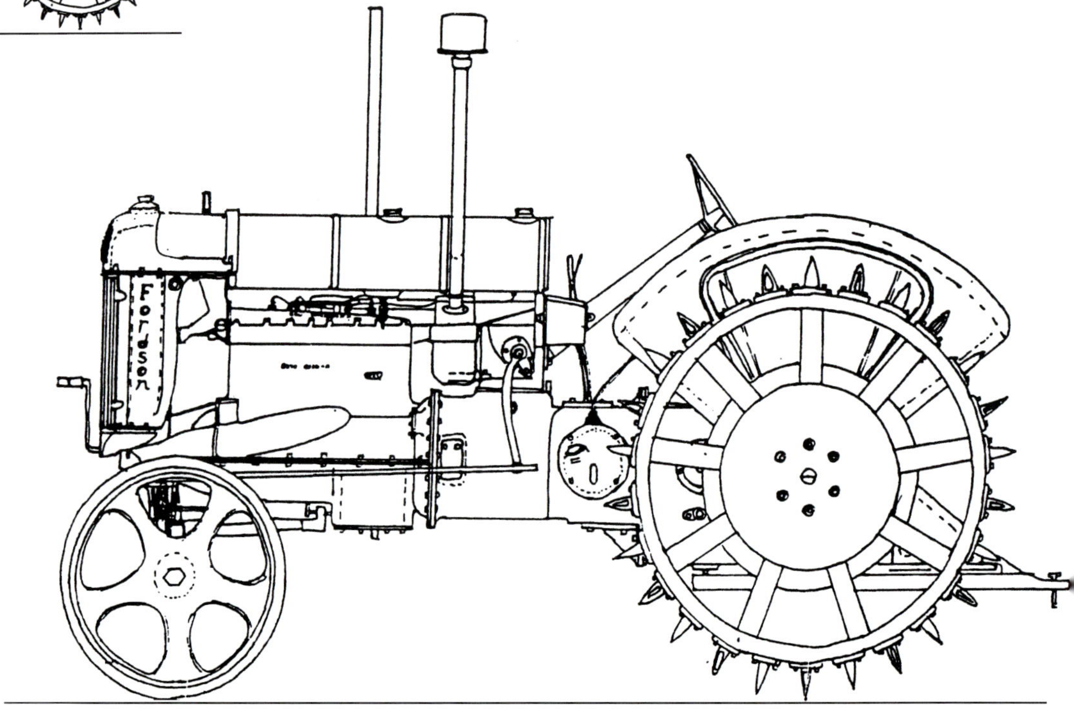

1945 EARLY PRODUCTION.

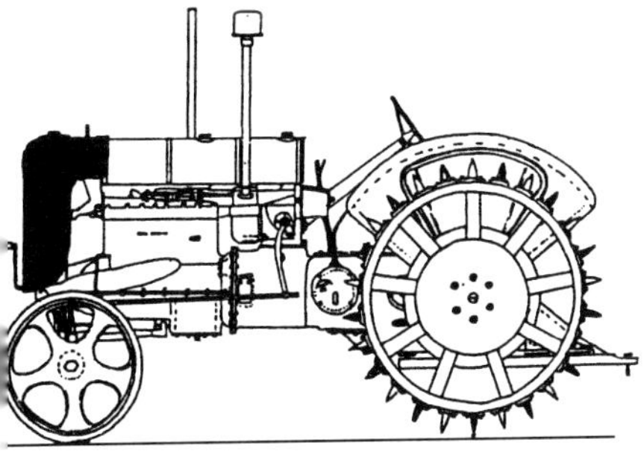

When production started, there was still a quantity of 'N' parts to be used up. The radiator design had still not been finalised, but on the first 100 or so tractors down the line the style of top tank shown here was used. Aircleaner, exhaust, sediment bulb, and toolbox were still all 'N' stock items. As the paint plant would not accommodate the new tractors, they were spray finished at the end of the truck line as long as 'N' assembly continued.

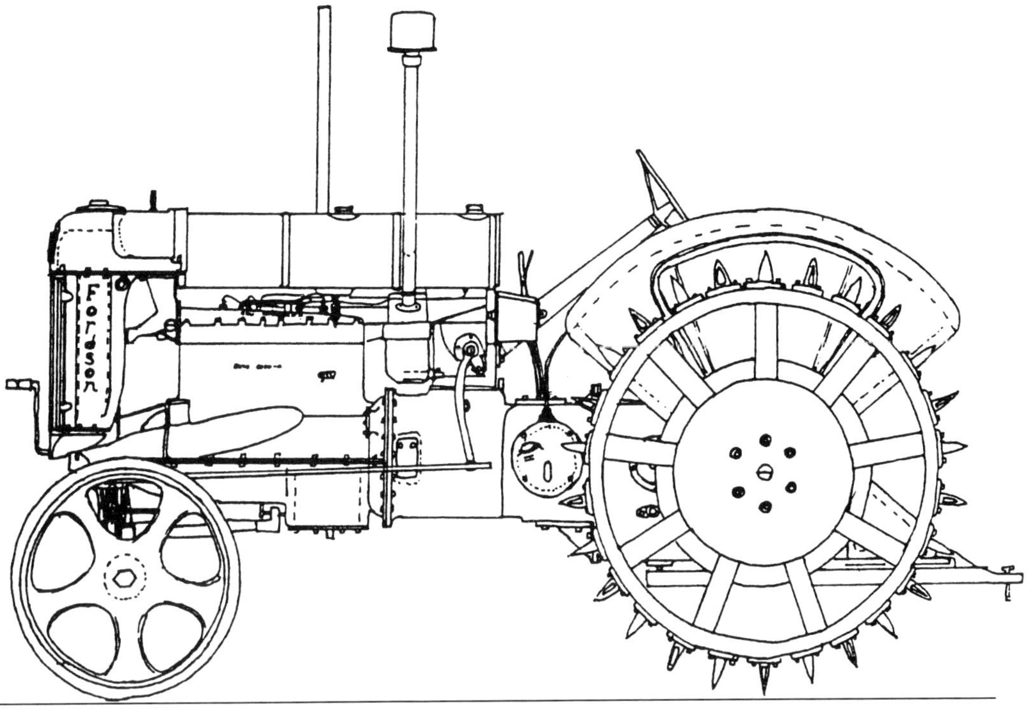

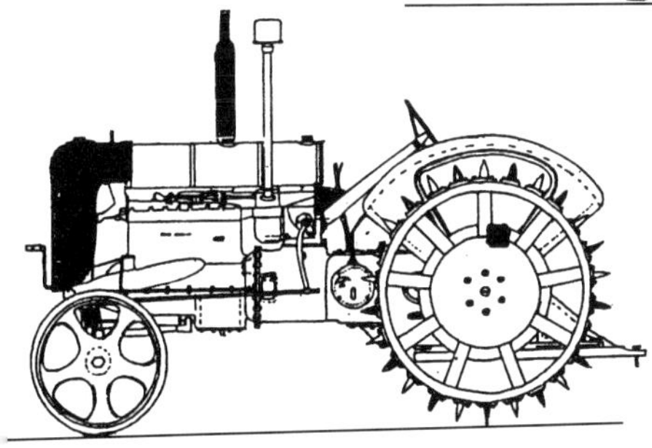

'N' production ended on 4th June 1945. By that time the radiator top tank design had been finalised, and the alloy cast grilles had come into use. The exhaust silencer had been replaced by a one piece 'Burgess' design from May, the stock of straight pipes and expansion boxes being used up on 'N' models and then transferred to the parts operation. The upper dash was altered in anticipation of future developments on the electrical front, and the toolbox, the large pattern, relocated under the seat. Due to supply difficulties from Willmot Breeden, an A-C pattern of pillar calormeter was used on some early tractors. The offset throttle linkage from governor to dash also vanished around this time. The large drawing shows a Row Crop tractor; the small one a Standard Agricultural.

JUNE 1945.

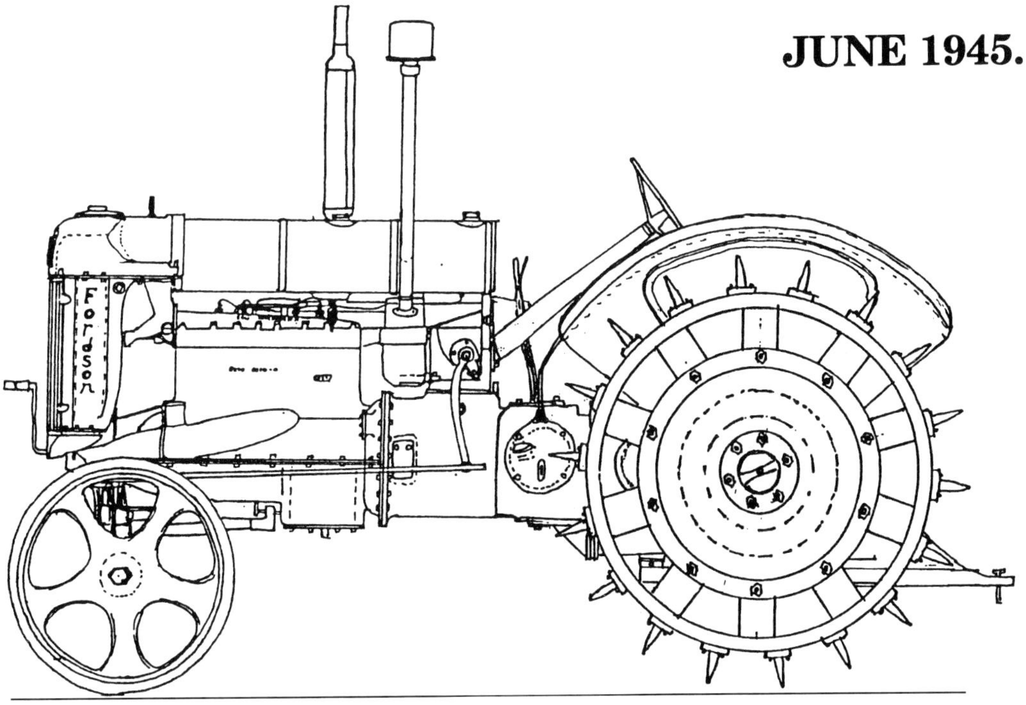

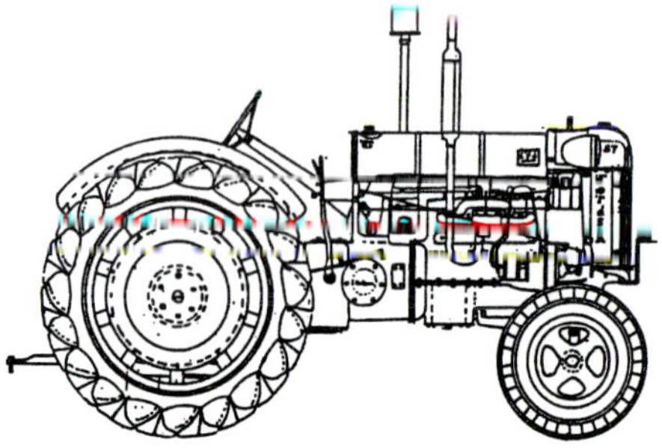

1945 LAND UTILITY

The first proper Land Utility tractors did not appear until June 1945. These had 6.00 x 19" front tyres and 9.00 x 36" rears on ten lug rims. If you wanted lighting on the E27N at this time the only option was to fit the six volt system as used on the model 'N' as shown here.

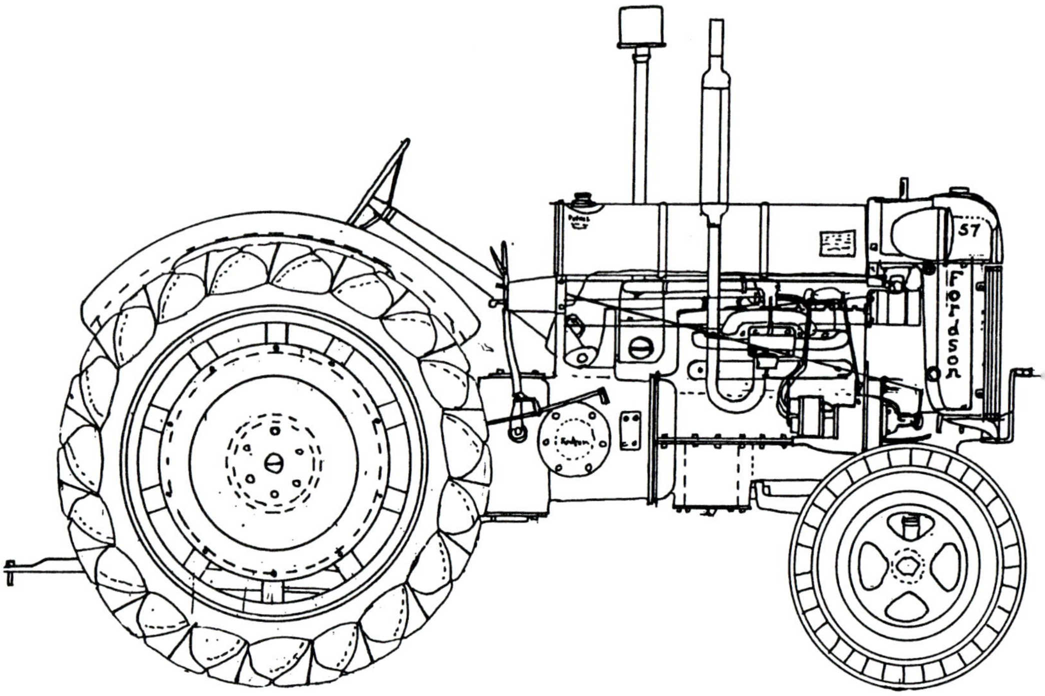

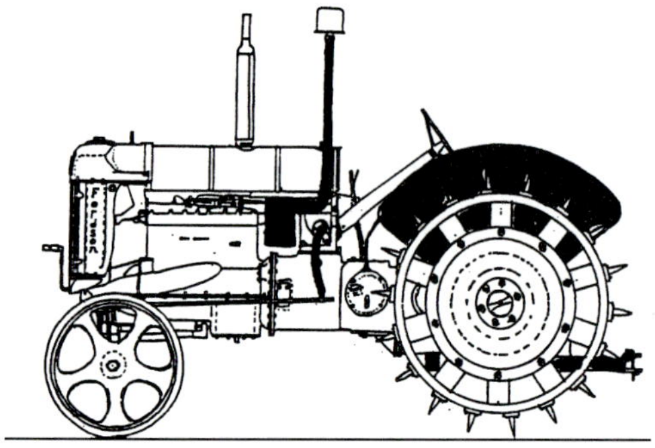

JANUARY 1946

By February 1946 the fixed level drawbar had gone, as had the square aircleaner, to be replaced by a standard 'Burgess' unit, the original pattern of radius rod had given way to a stronger, braced, pattern, and a new type of mudguard was used. A Standard Agricultural tractor is shown with the Row Crop equivalent in the small drawing.

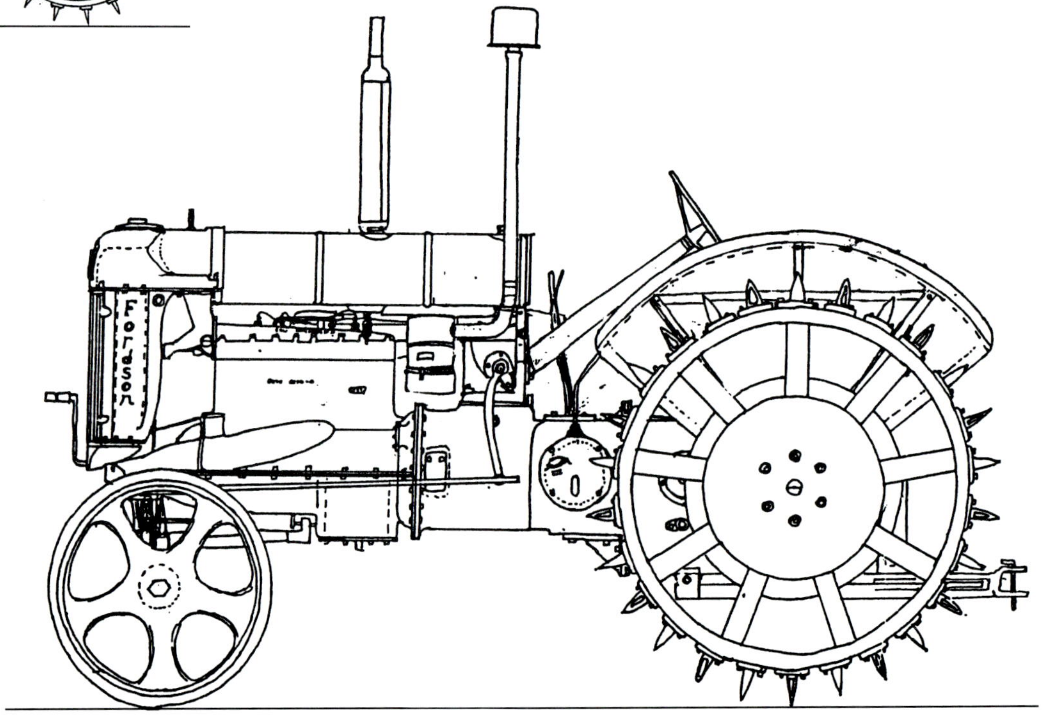

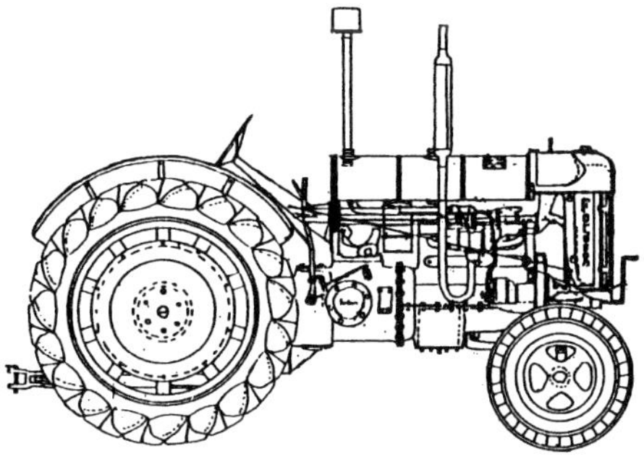

1946 LAND UTILITY

This drawing shows the equivalent 1946 Land Utility model. Lucas GJ4 magnetos were used on all E27Ns, and on this example dating from March 1946, a starter ring gear would be fitted. The upper dash designed to accommodate the battery for the new 12 volt electrics is now fitted. 9"rear tyres are still used, however.

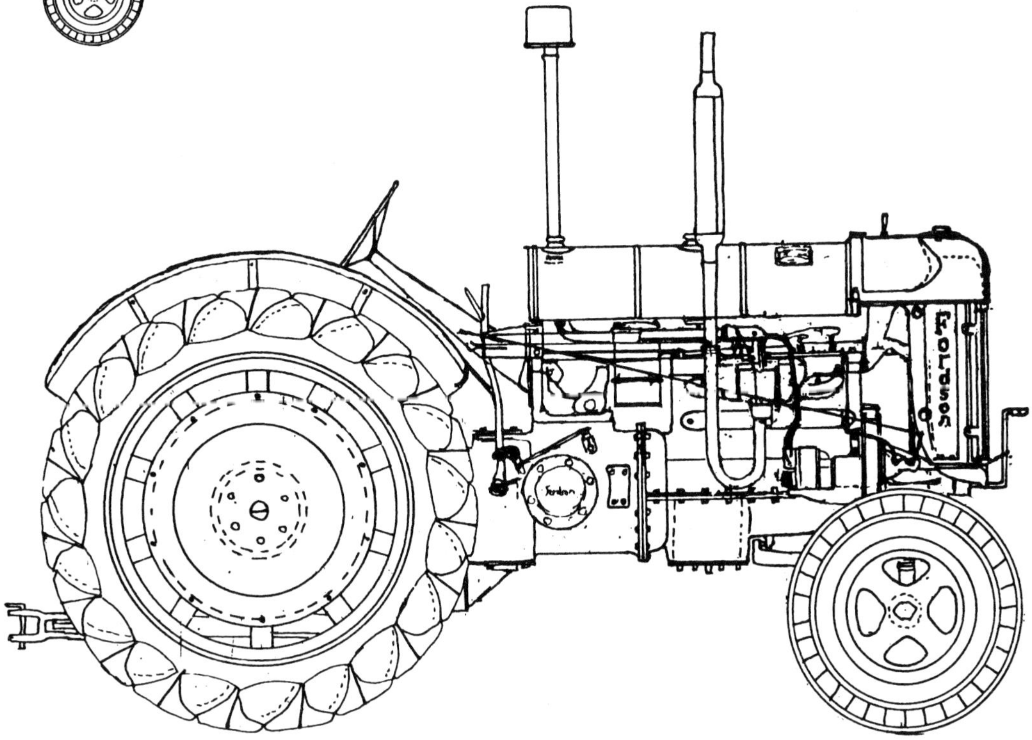

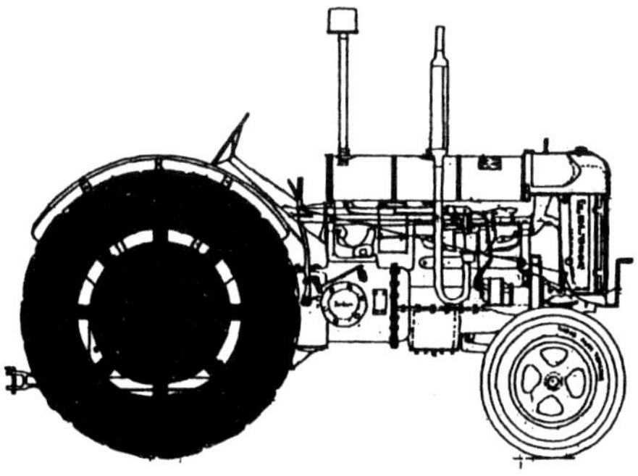

1947 LAND UTILITY

By early 1947 the rear tyre specification had been amended to 11x36" cases, and the wheel rim had only 8 locating lugs.

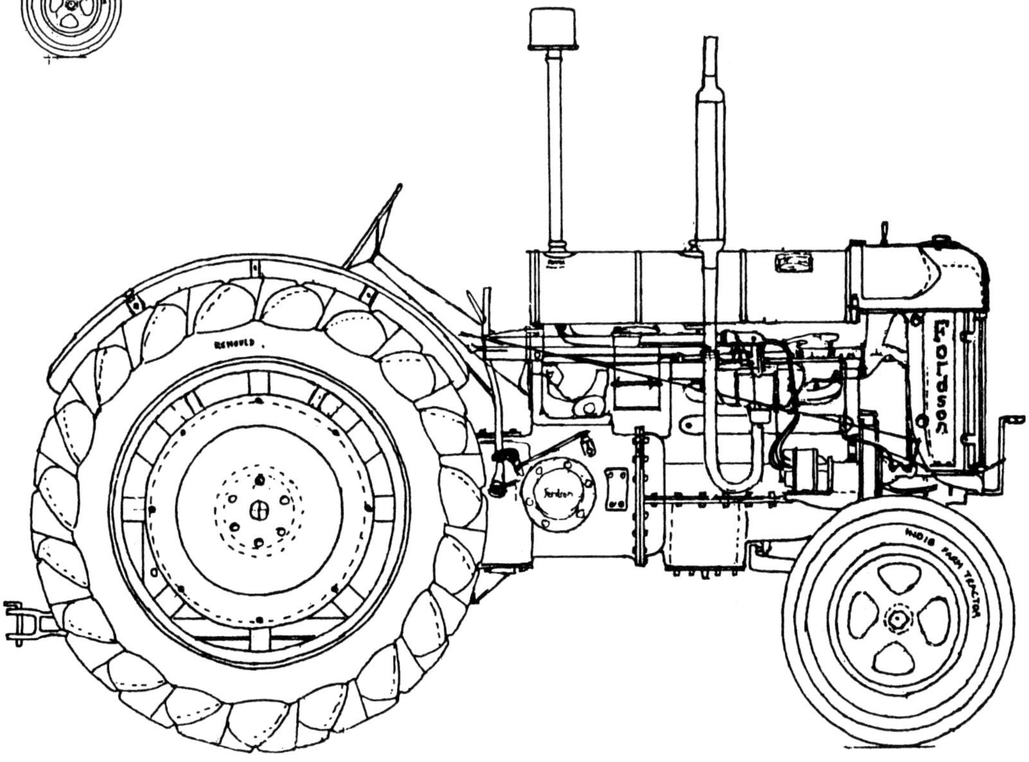

MID 1947 LAND UTILITY

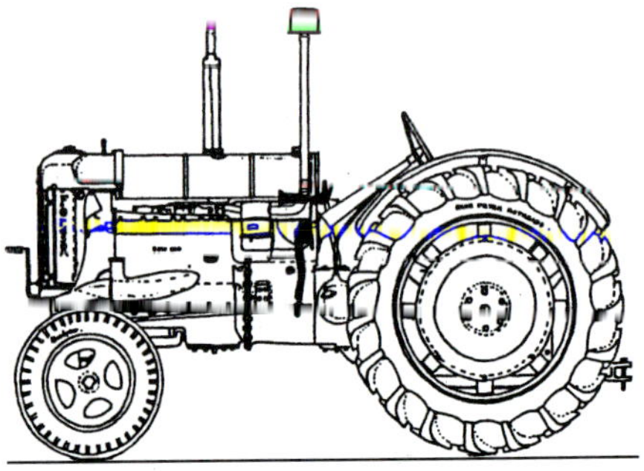

This mid-1947 tractor, seen from the nearside, shows the adoption of the worm and nut type steering and the return to a straight drop arm. Many tractors from 1946-7 had a cranked drop arm designed to clear certain toolbar equipment when fitted. The Lucas No. 12 bulb horn is also prominent; it was fitted to all LU and RC tractors on pneumatics less lighting and/or starting.

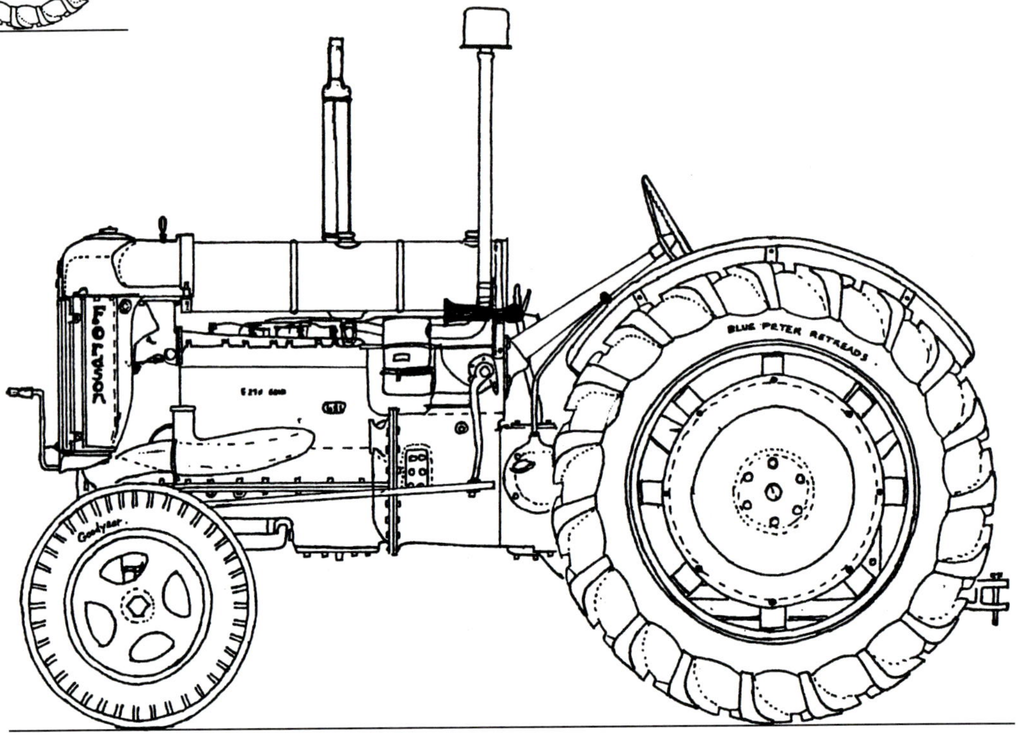

LATE 1947 ROWCROP

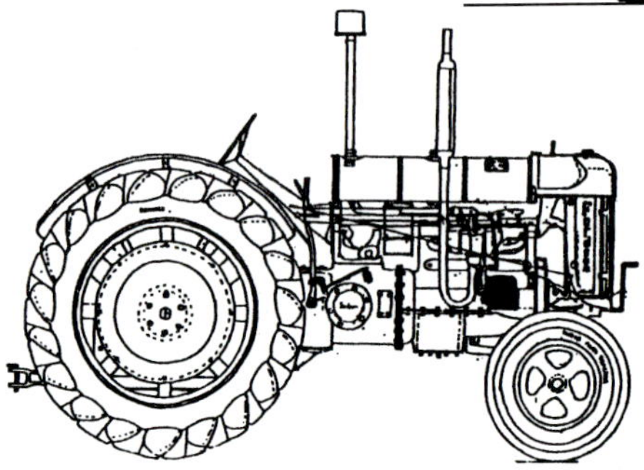

In late 1947 Lucas discontinued production of their GJ4 magneto and introduced their 4RF 'Marine' pattern, seen fitted here to a Rowcrop tractor.

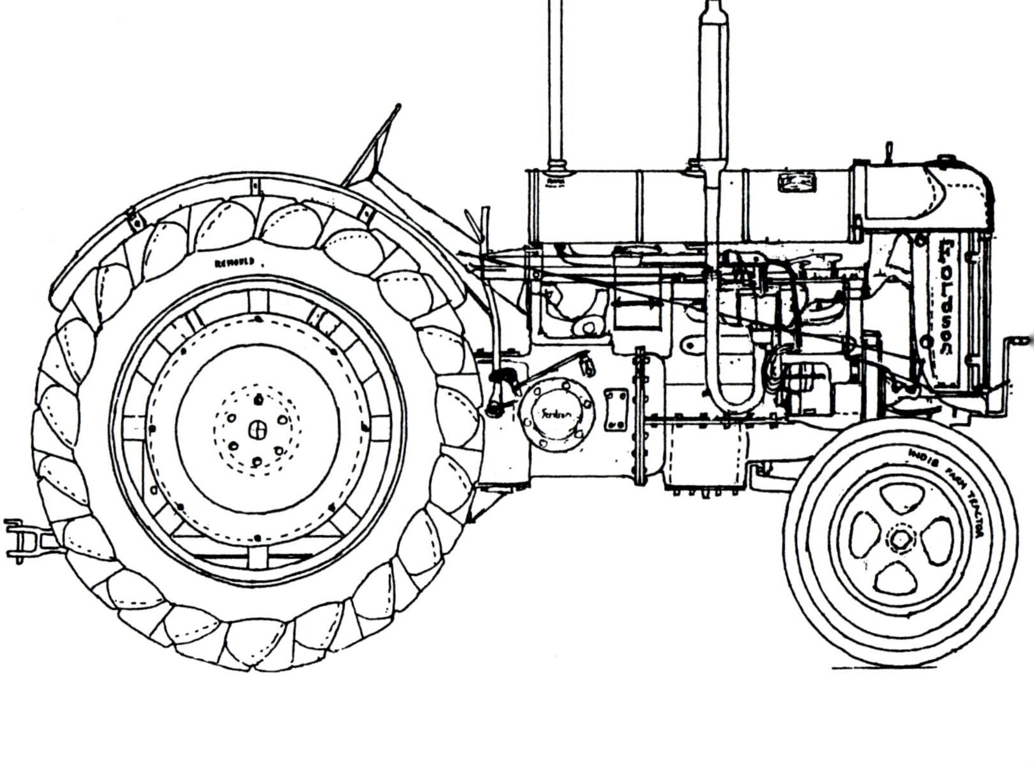

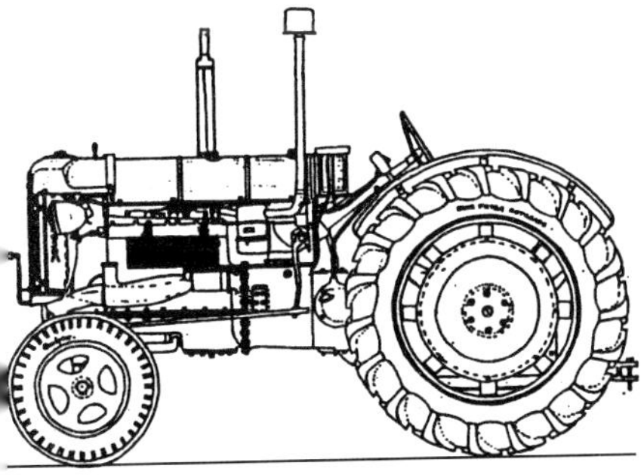

MID 1948 LAND UTILITY

A 1948 Land Utility is seen here fitted with starting and lighting. The wiring conduit was originally brazed to the tank. The toolbox is now mounted on the side of the engine, the normal location when hydraulics are fitted; this location became standard for all tractors in 1948.

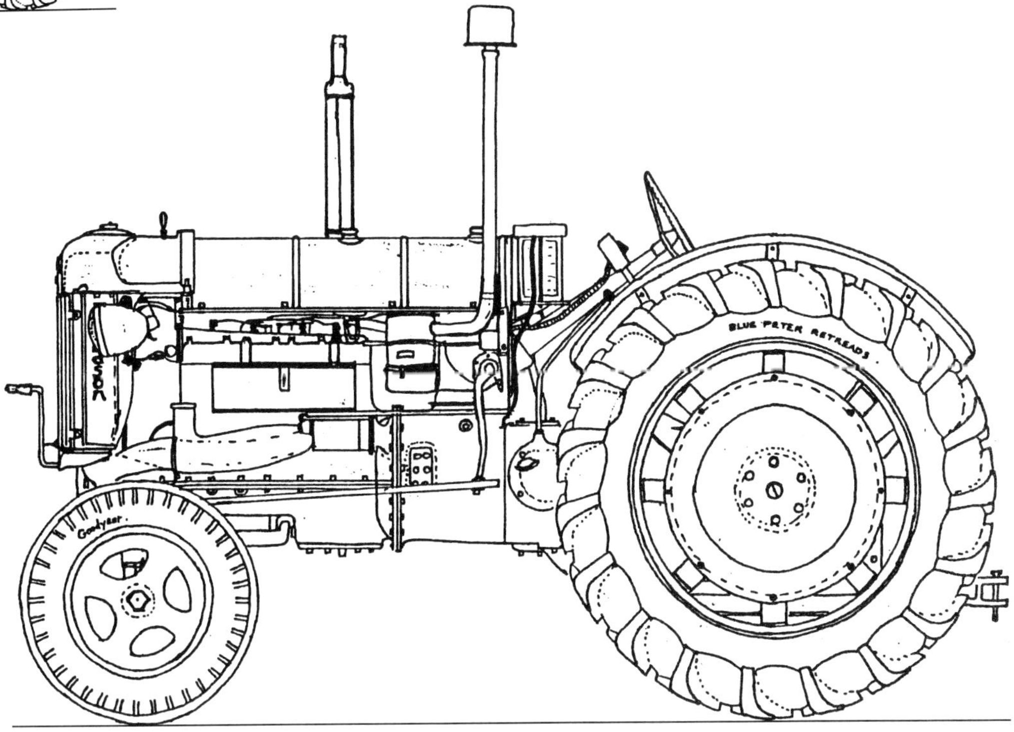

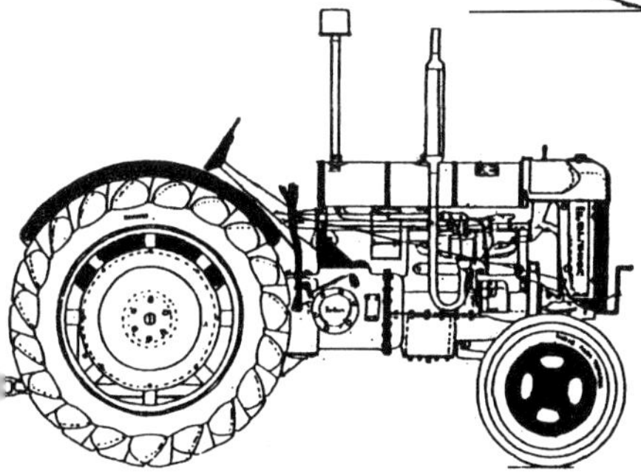

LATE 1948/1949 ROWCROP.

By early 1949 the mudguard design had again changed, the lower dash casting patterns had been remade to eliminate the mounting for the old style aircleaner, alloy steering wheels were now the norm, as were pressed steel handbrake levers. The front wheel castings now incorporated a dirt ring on the inside, and the grille introduced for the diesel with narrow bars came into use on all tractors.

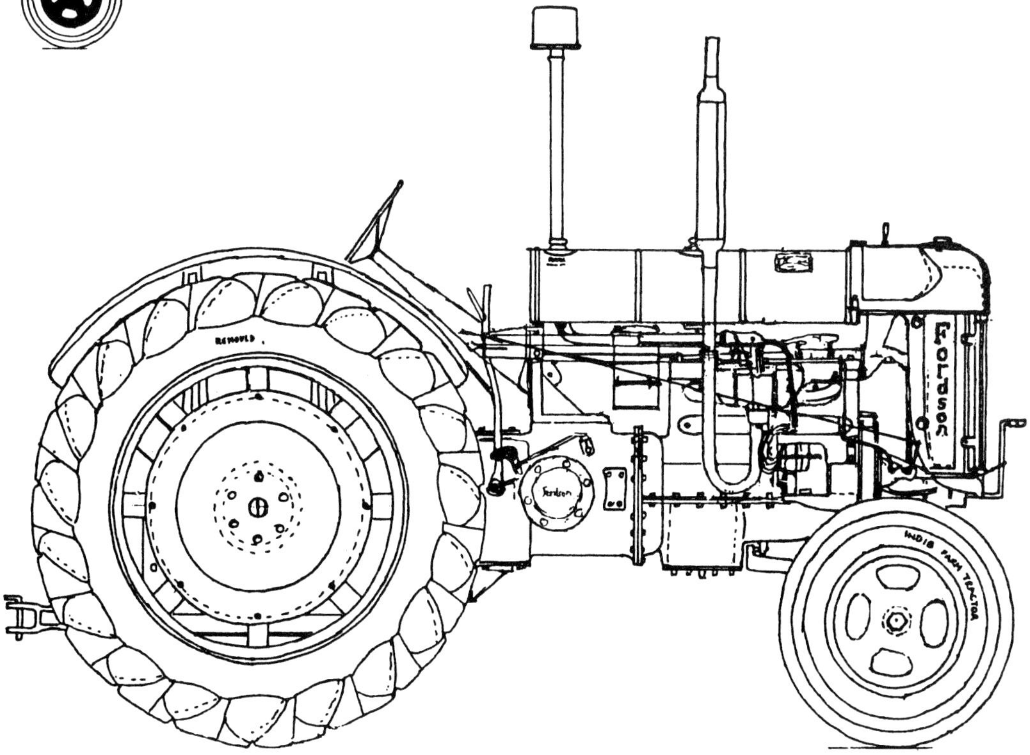

1948 P6(TA

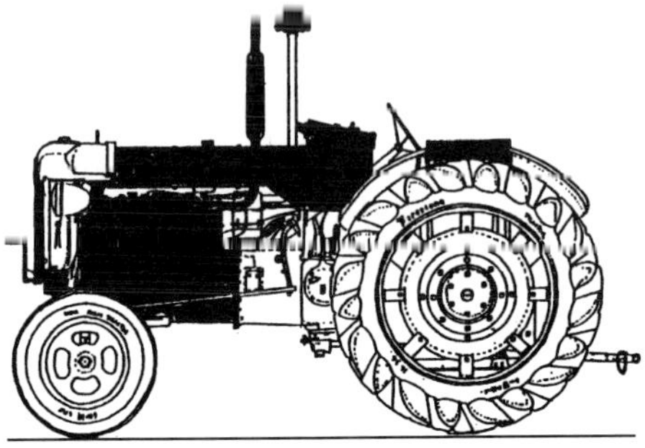

In late 1948 the diesel model with Perkins P6(TA) engine came int production. The black areas on the small drawing show parts unique to th diesel model.

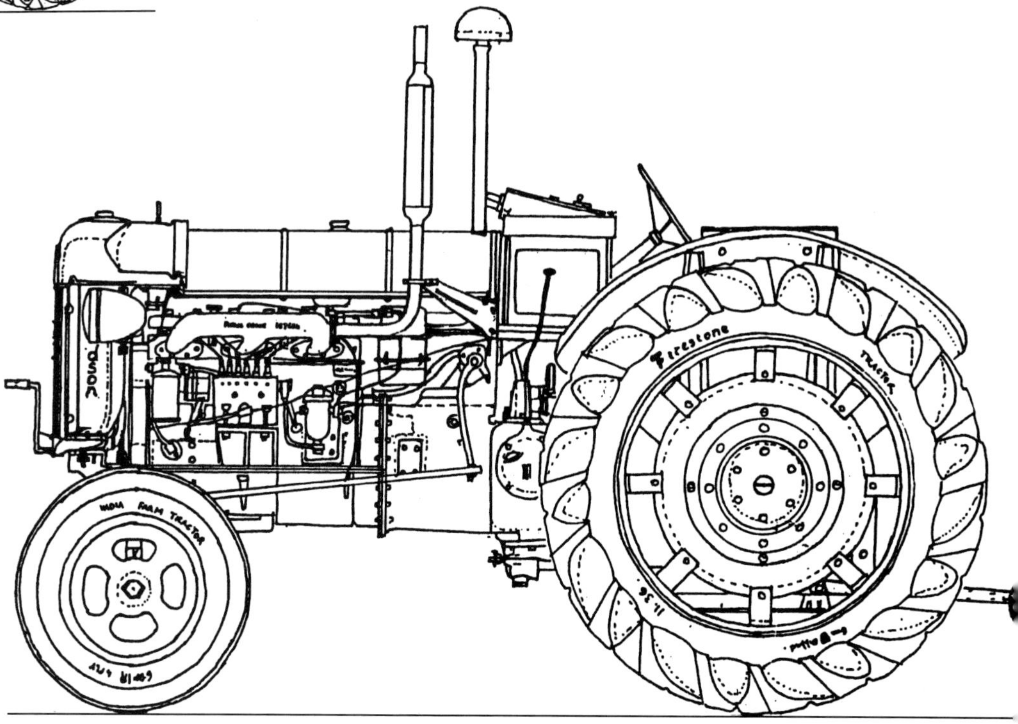

1950 P6(TA

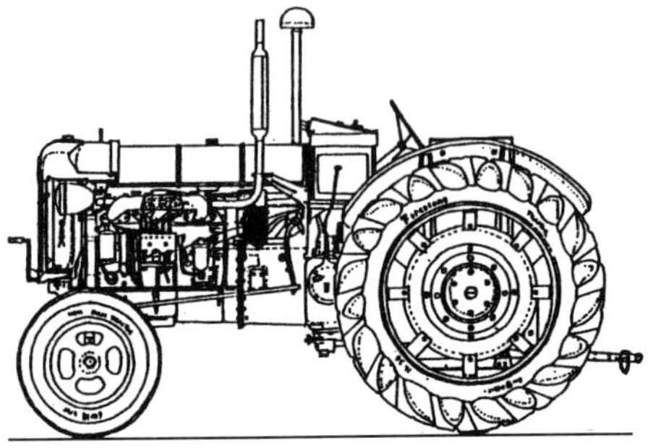

Late diesel models and subsequent conversions had an improved engine wit modified breathing and fuel filtration systems.

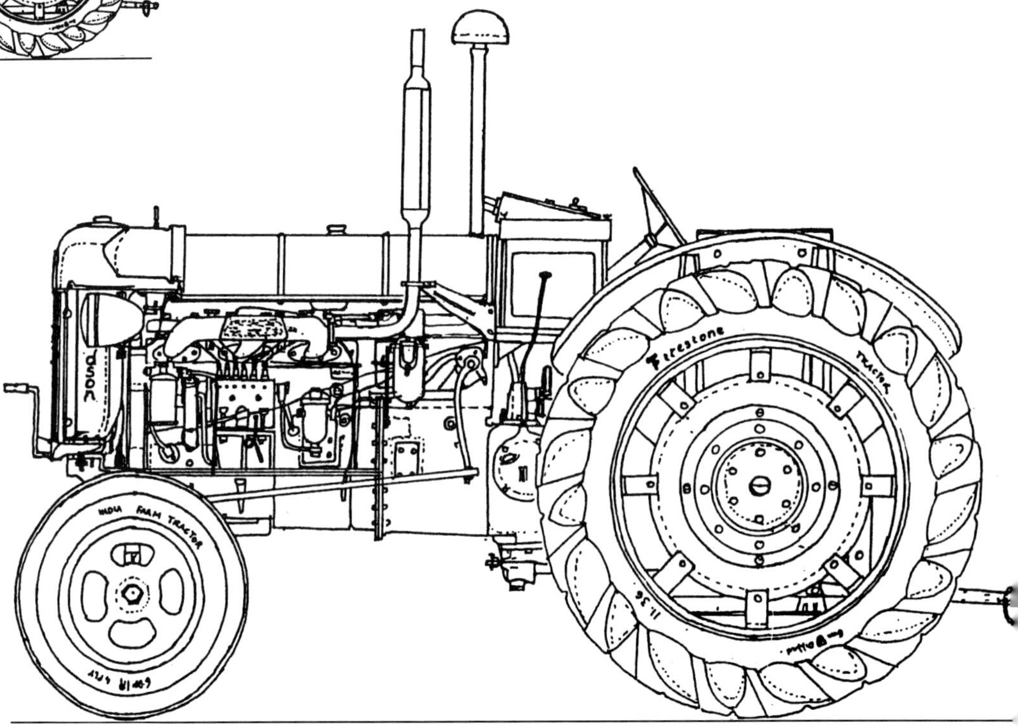

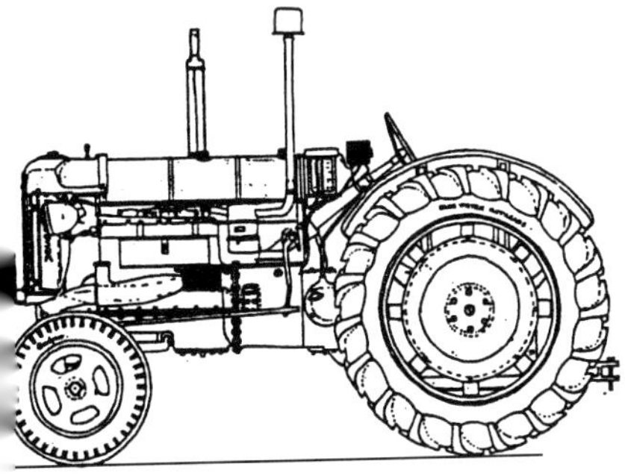

1949/50 LAND UTILITY

March 1949 also saw alterations to the electrical system with the use of a direct rather than a solenoid controlled starter with the relocation of the switch in a new style oblong control box, which also incorporated the voltage regulator. The wiring conduit now clipped between the front tank strap and the upper dash. Six bladed fans were also usual at this time.

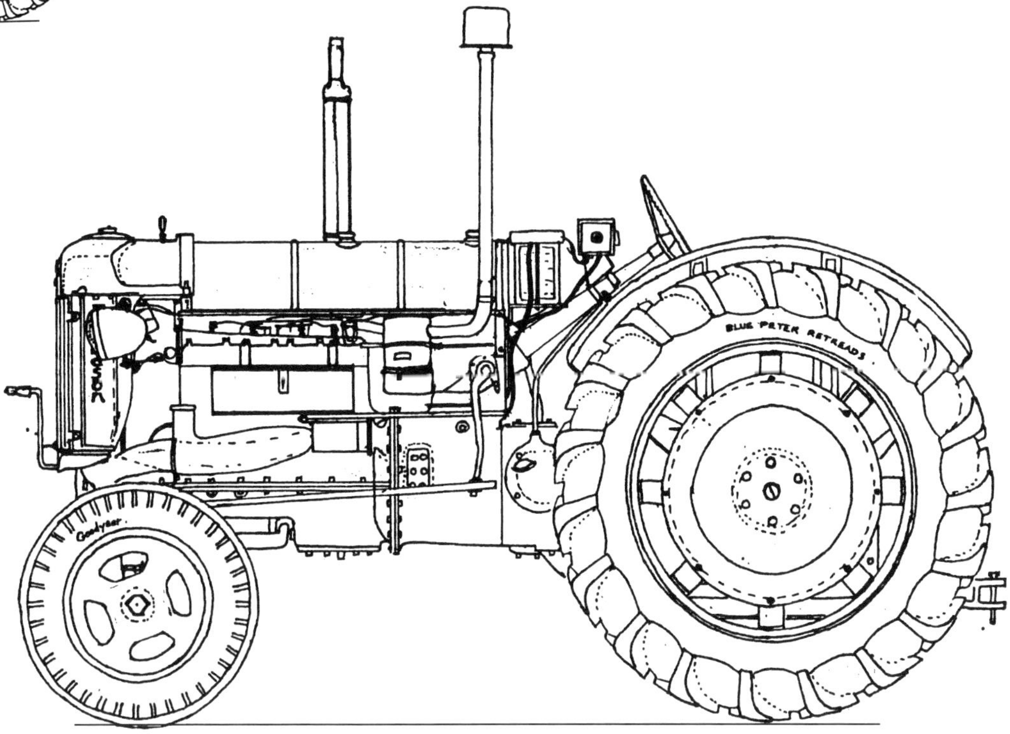

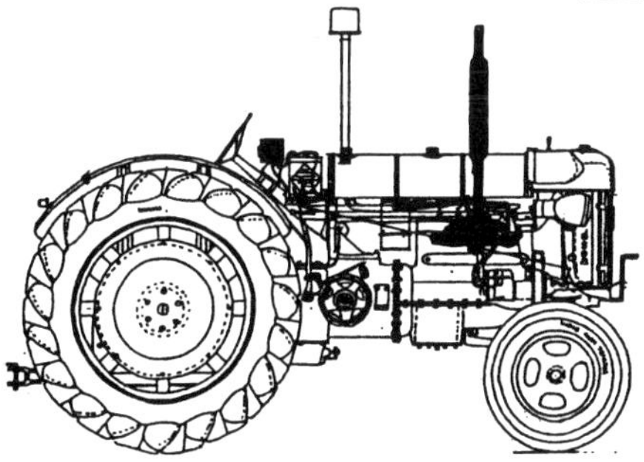

MID 1950-52 ROWCROP.

From 15th May 1950 the improved vaporiser was fitted in production. This required a modified silencer, and ignition conduit. The tractor shown also has the A-C Sphinx aircleaner rather than the more common Burgess type.

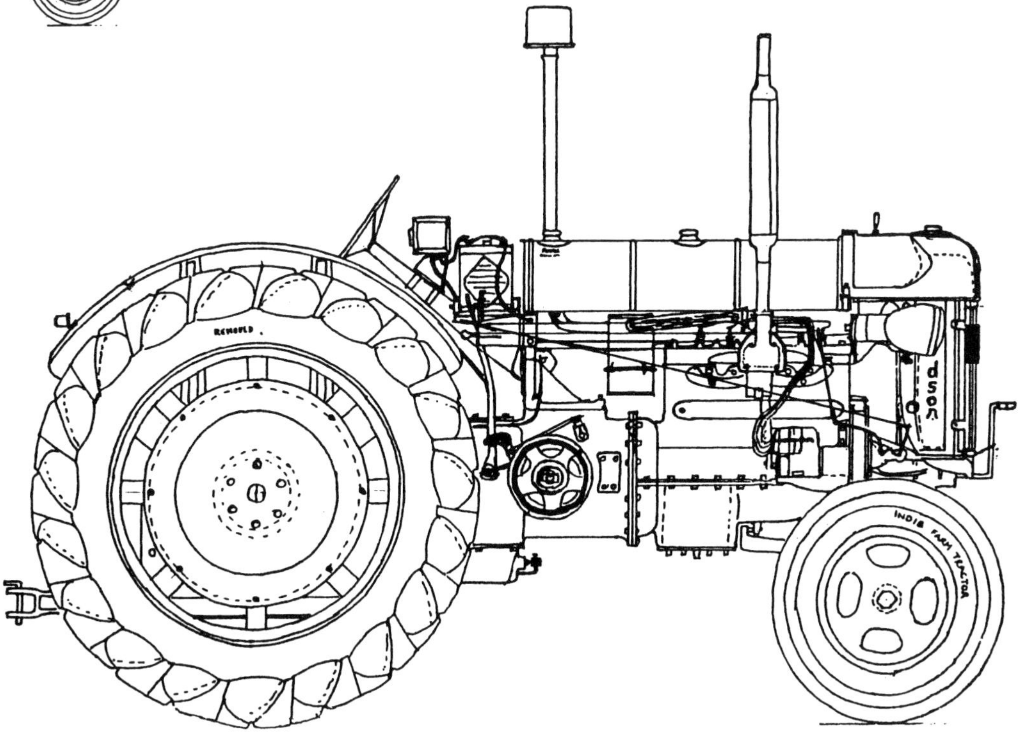

INDUSTRIAL

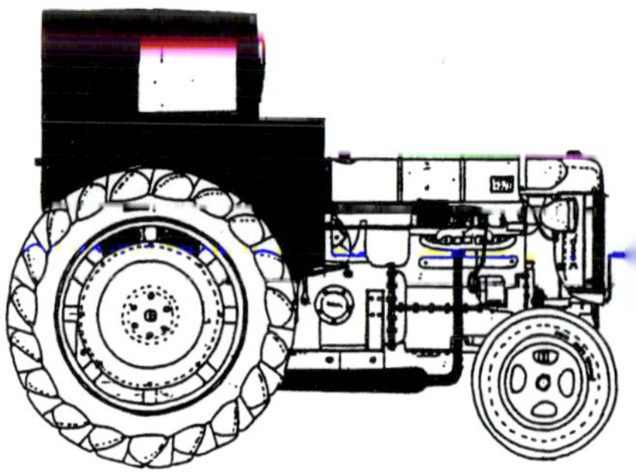

1947 saw the introduction of the industrial model and parts unique to this are shown in black on the small drawing.

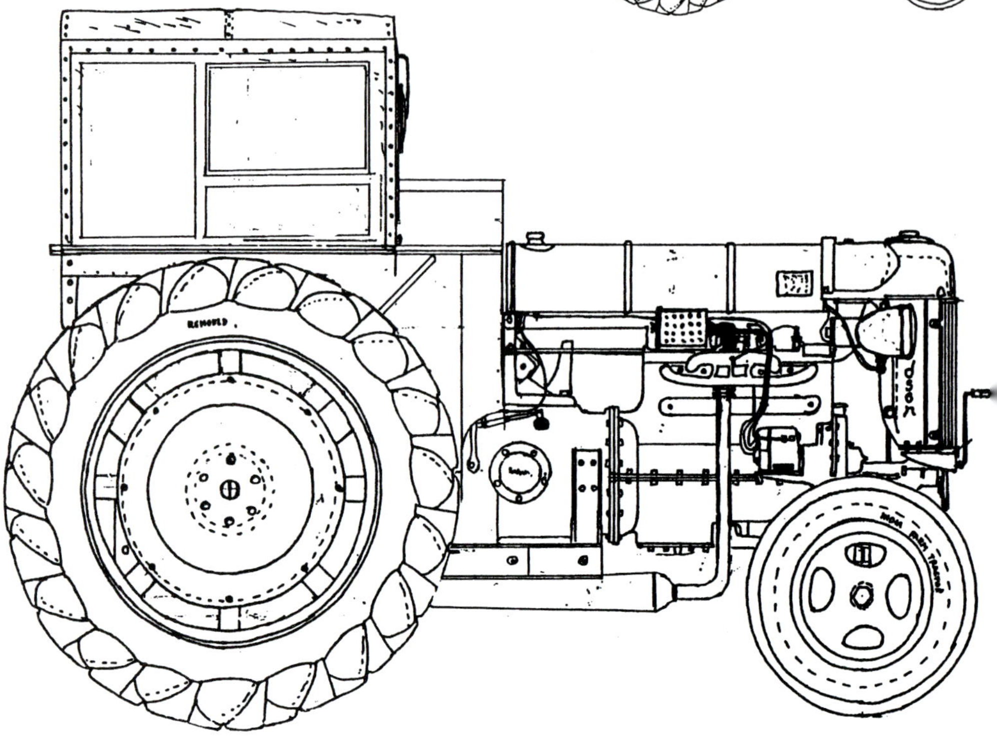

DG4 HALF TRACKS.

Roadless DG4 tracks were fitted to E27Ns from almost the start of production. The tractor shown however is of more interest as being a gas oil equipped unit for export - the glass jar type pre-cleaner was fitted for some territories. Note also the pressure cap installation on the radiator.

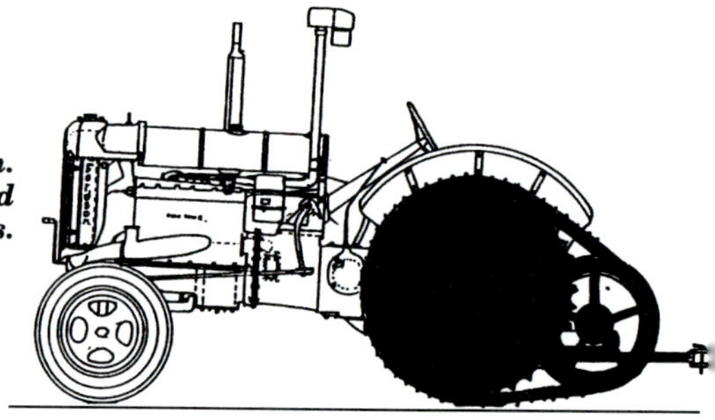

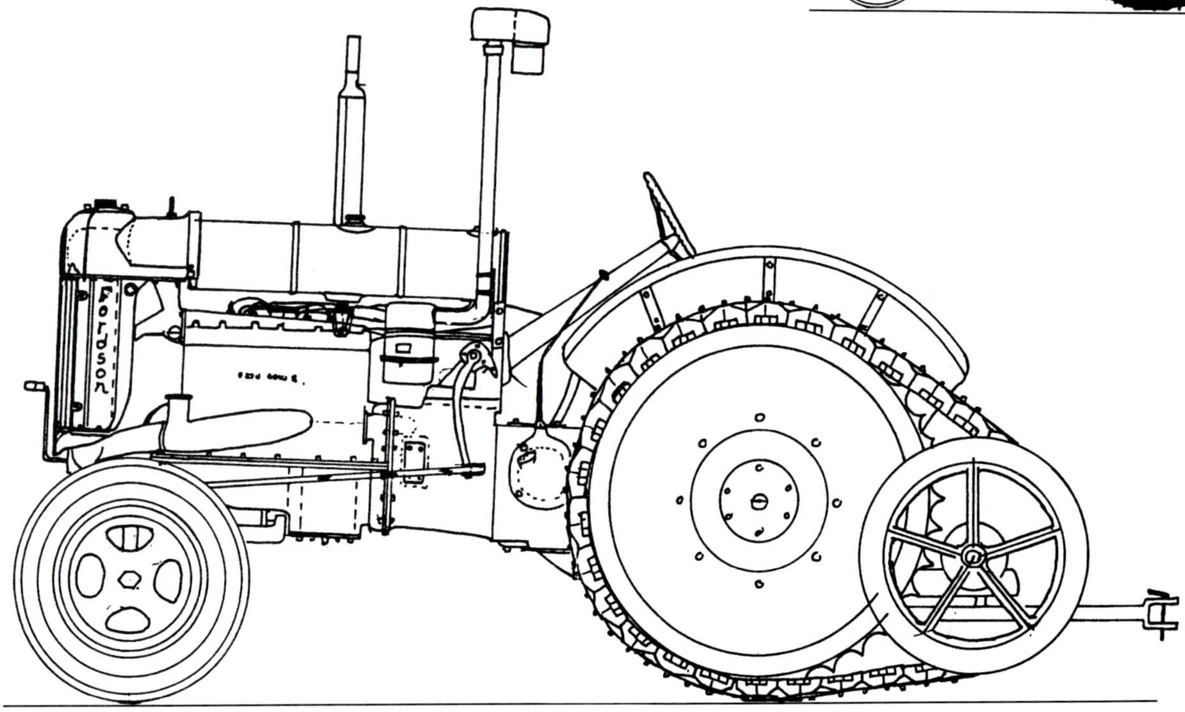

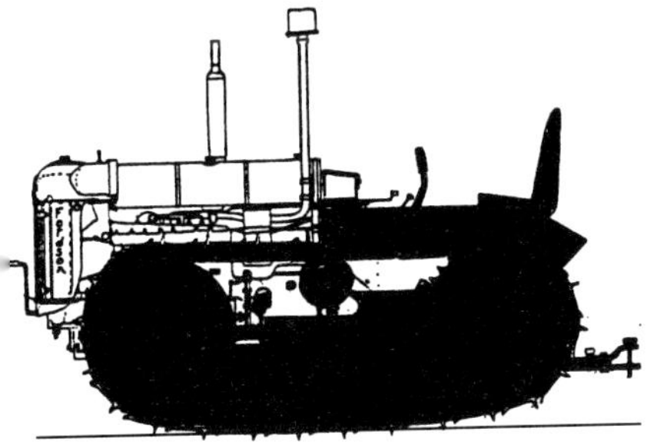

COUNTY FULL TRACKS

The County Crawler was developed in conjunction with Ford Motor Company. It used a modified E27N rear axle with solid drive instead of a differential. The spark ignition model is shown here, and components unique to the model are outlined in the small drawing.

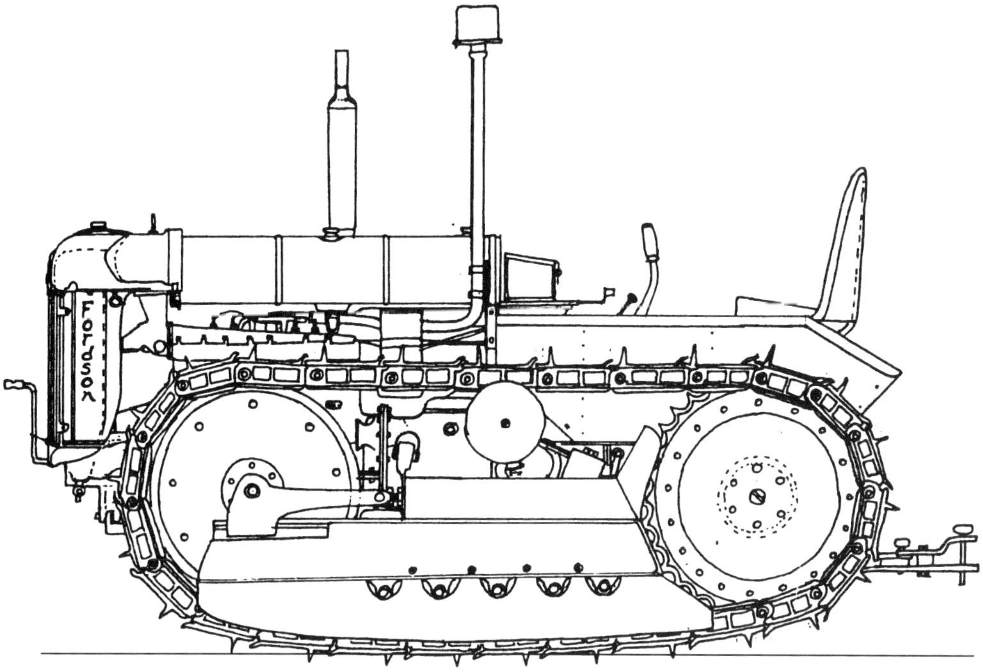

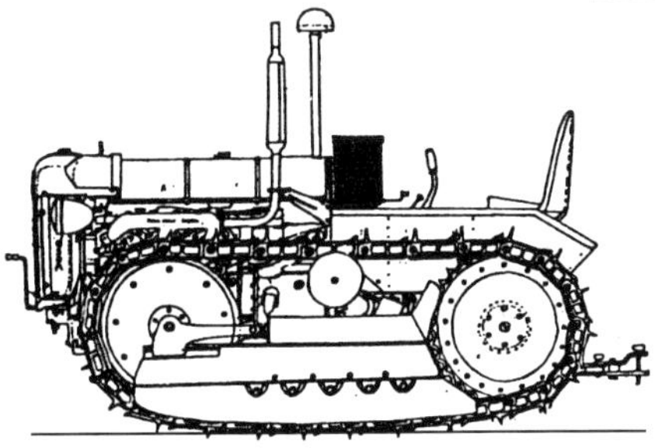

COUNTY P6(TA)

The P6(TA) engined model County is seen here. Special battery carriers were fitted to this model, and the control box located by the LH mudguard.

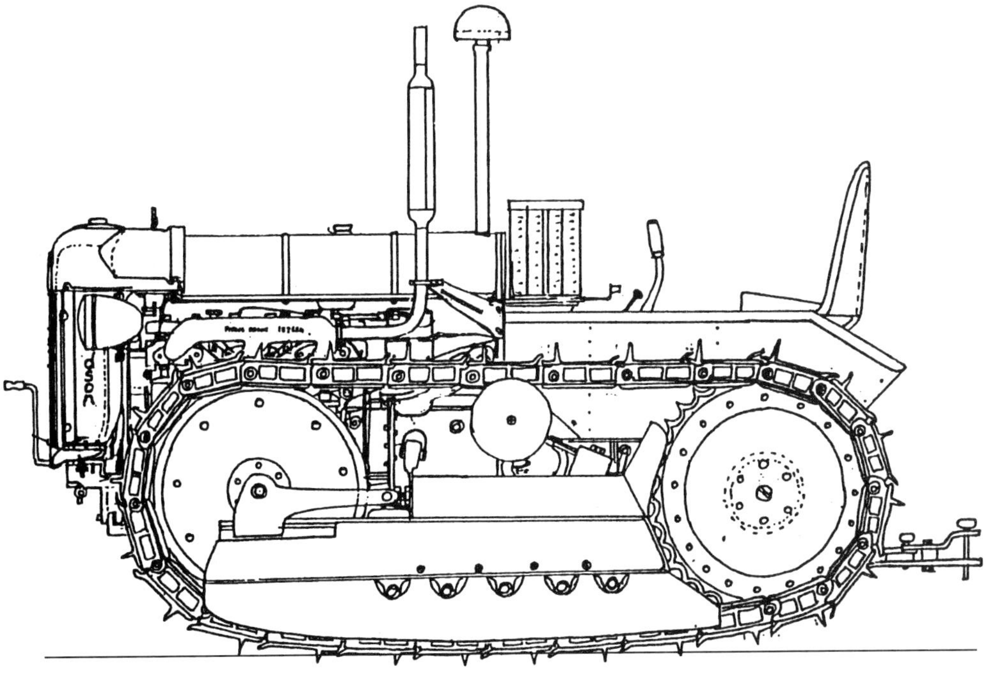

P4(TA) PROTOTYPE

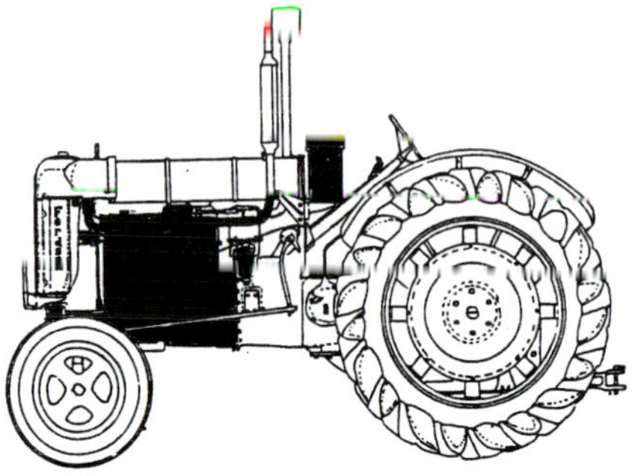

Before the P6(TA) was adopted in production a P4 engined tractor was built by Perkins for evaluation. The drawing here shows the installation which required the shortening of the fuel tank.

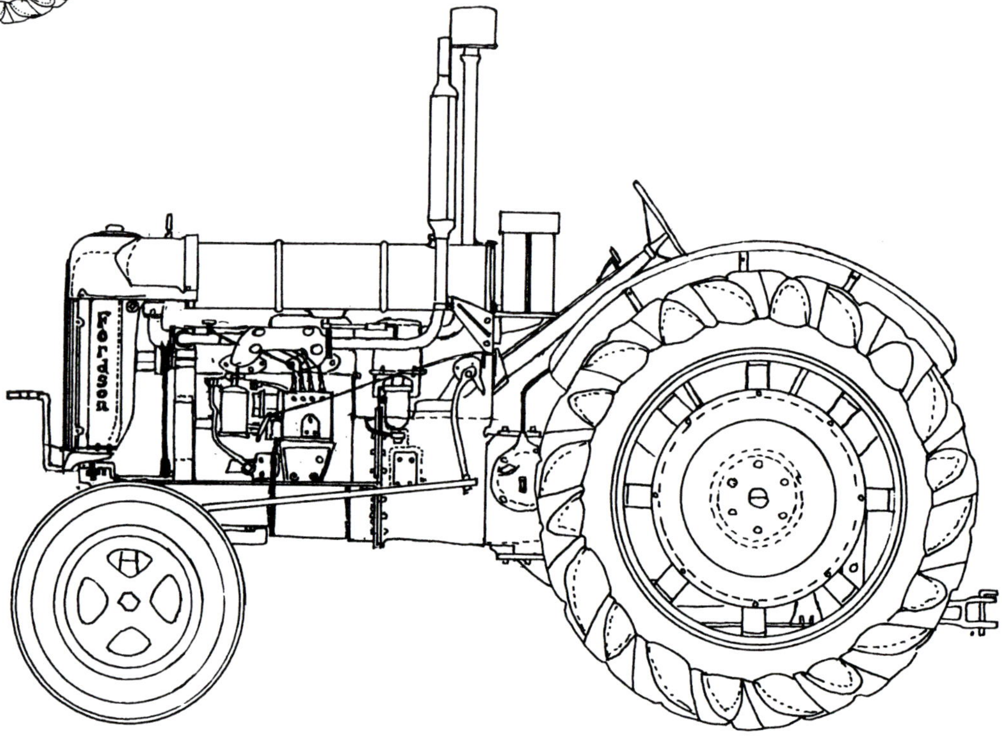

L4(TA) CONVERSION

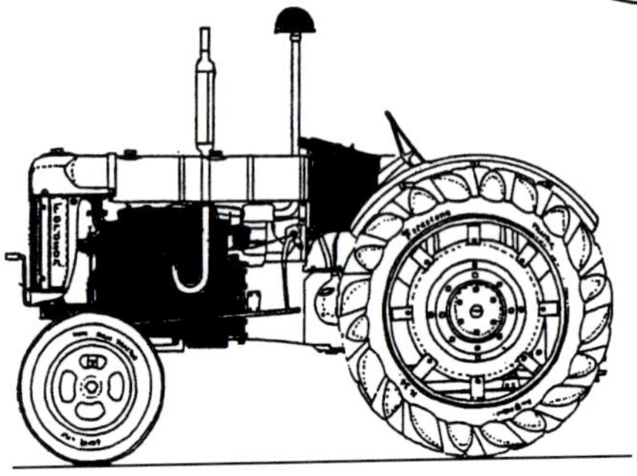

From 1953, many Perkins L4 conversion packs were sold to convert S.I. tractors to diesel, and the components unique to this conversion are shown in black in the small drawing.

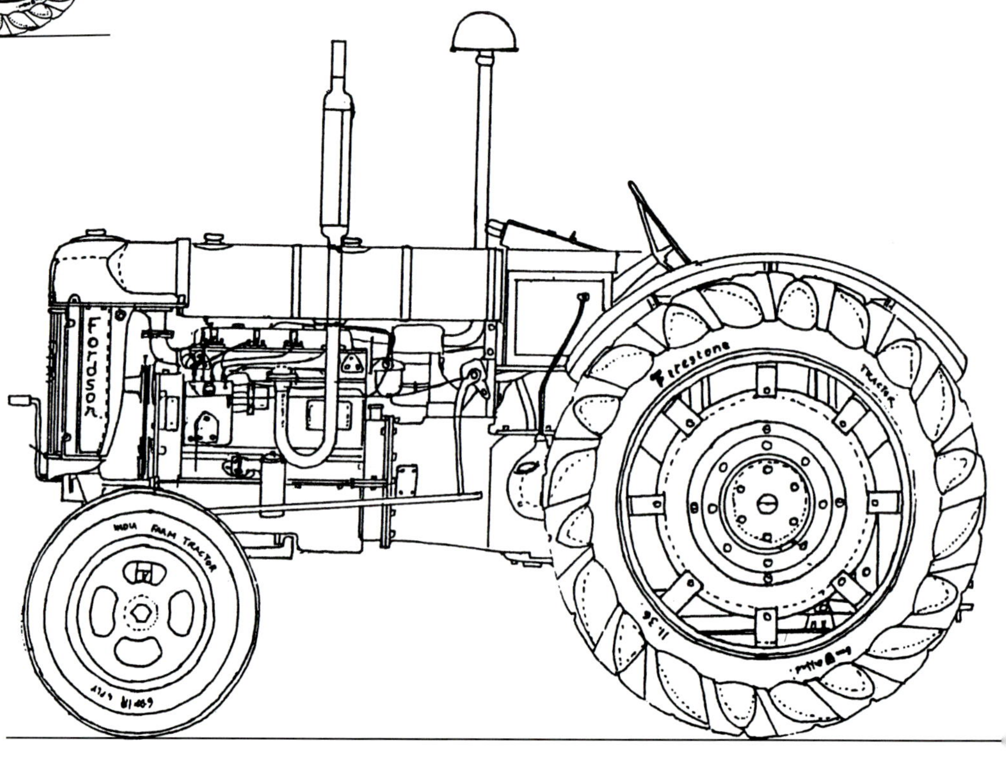

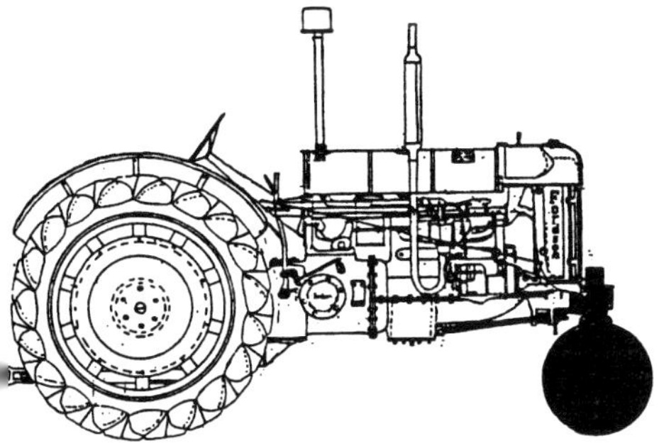

BETTINSON SINGLE FRONT WHEEL.

Ford never produced a tricycle E27N themselves, but a number of conversions were undertaken by Bettinson of Holbeach. One is seen here.

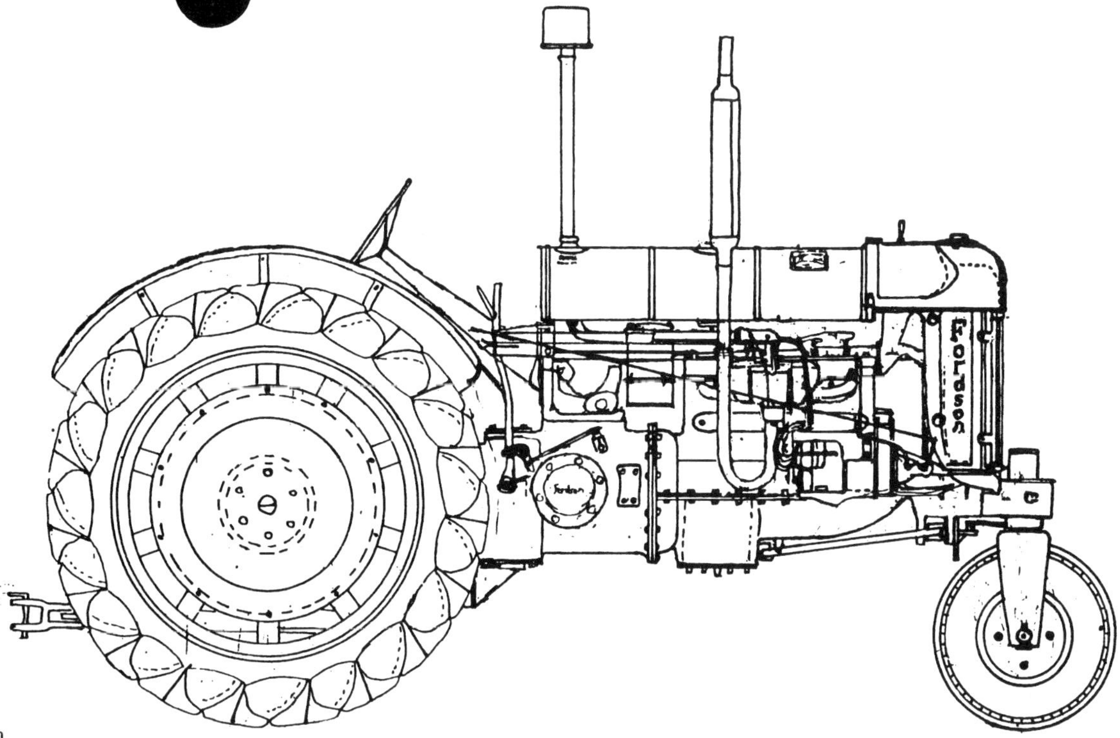

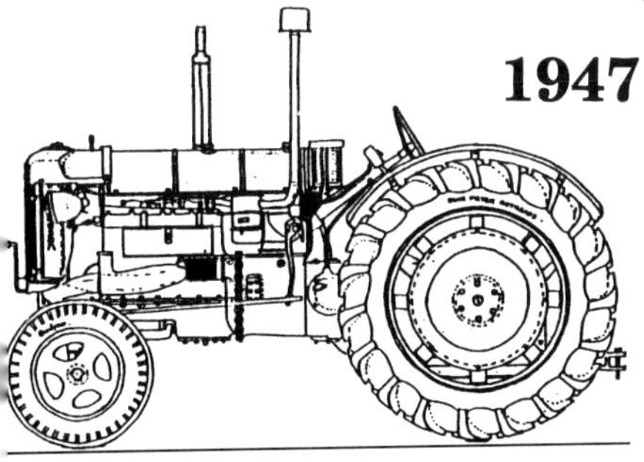

1947 LU. CONVERTED TO DIRECT STARTER.

If a direct starter had to be fitted in service to a tractor built prior to March 1949, a bracket to hold the new switch had to be fabricated and fixed to the dash as shown here. This detail will assist readers who are in difficulties in getting the correct parts to restore a pre 1949 tractor.

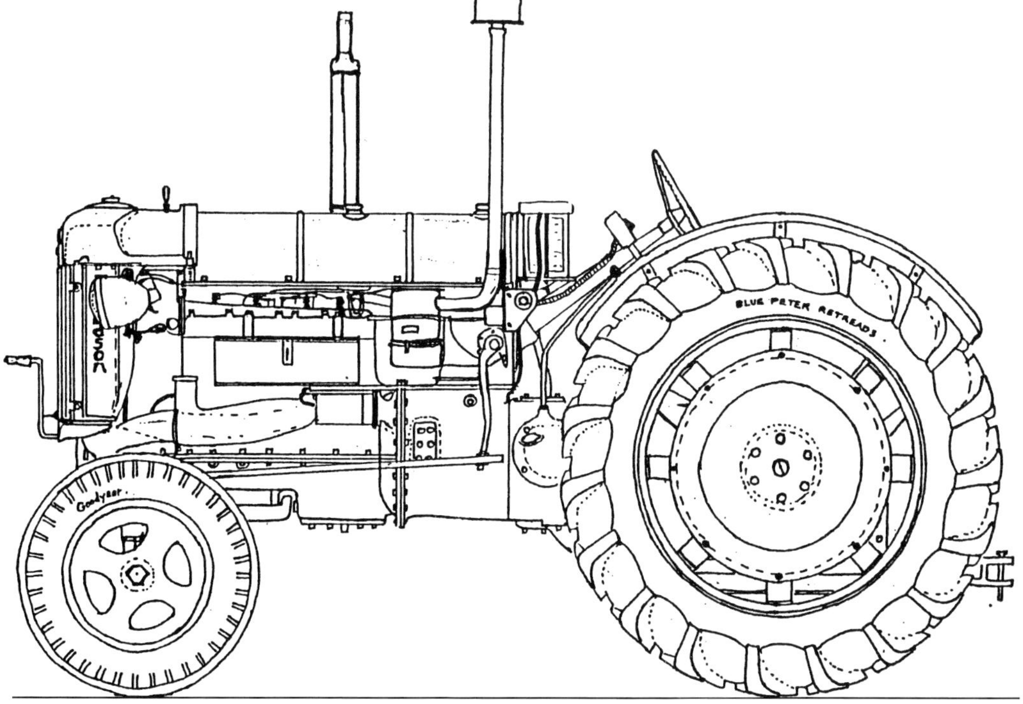

1949 AUSTRALIA

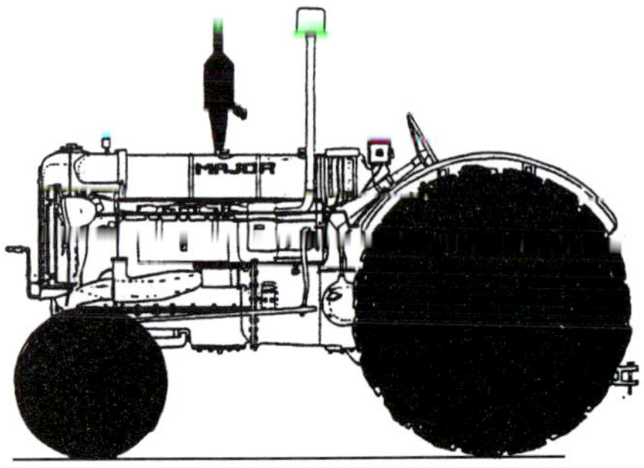

A 1949 Australian model is seen here, complete with Yuba spark arrester, radiator shutters, pressure cap installation, wheel centres & tyres to suit the Australian market, and 'Major' transfers on the tank.

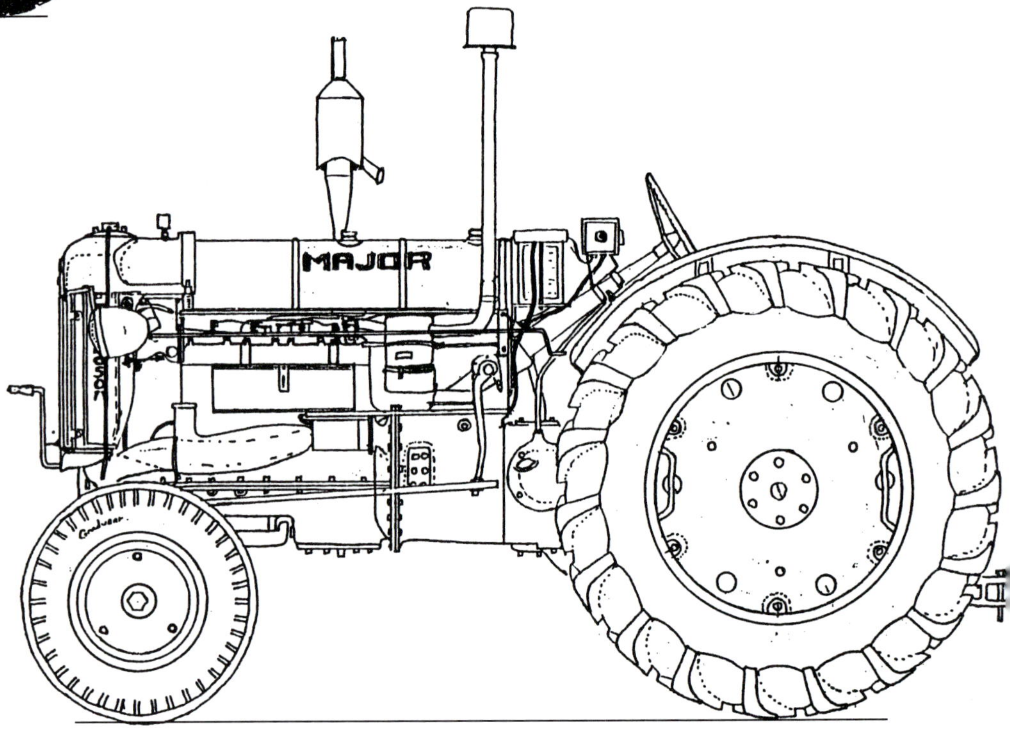

1950 AUSTRALIA

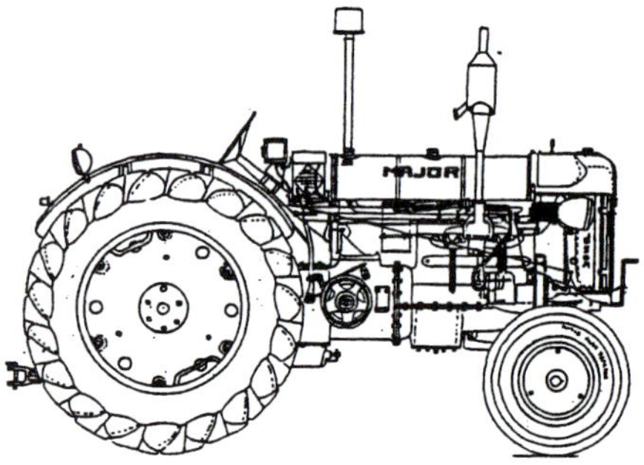

The offside view of a 1950 tractor sold in Australia. The dynamo is driven by a separate belt from the fan pulley, the manifold has extra long drain valve deflectors, and other features seen in the previous illustration are evident.

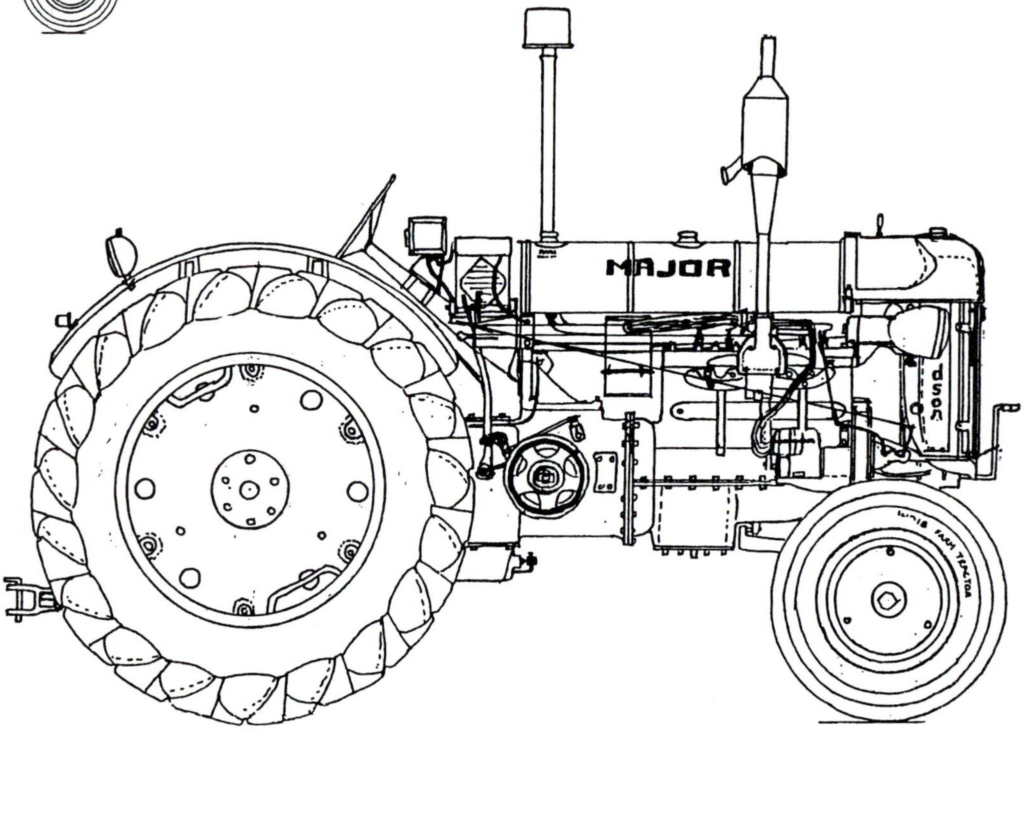

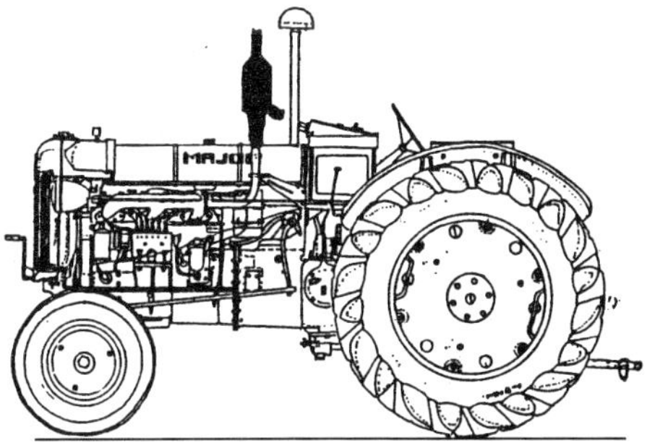

1949 AUSTRALIA P6(TA).

Sales of E27Ns in Australia went on well into 1953. Here we see a P6(TA) engined example with all the unique features fitted to tractors 'down under'.

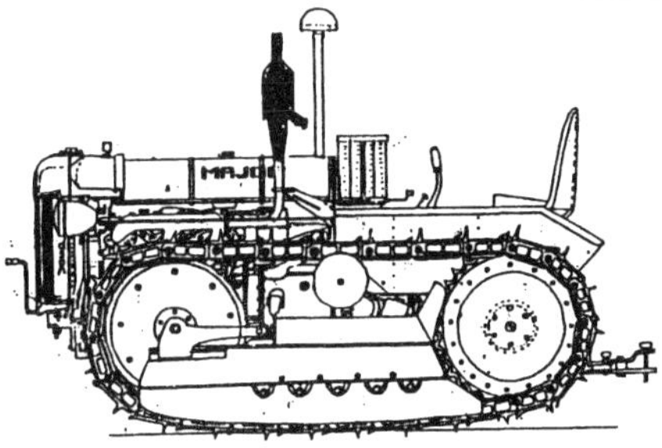

1949 AUSTRALIA COUNTY P6(TA)

County Crawlers were also sold in Australia, and had the usual extras added before sale. Note the extension of the overflow pipe right down to the base of the radiator, equipment fitted in the UK had only a short pipe which terminated just below the top tank.

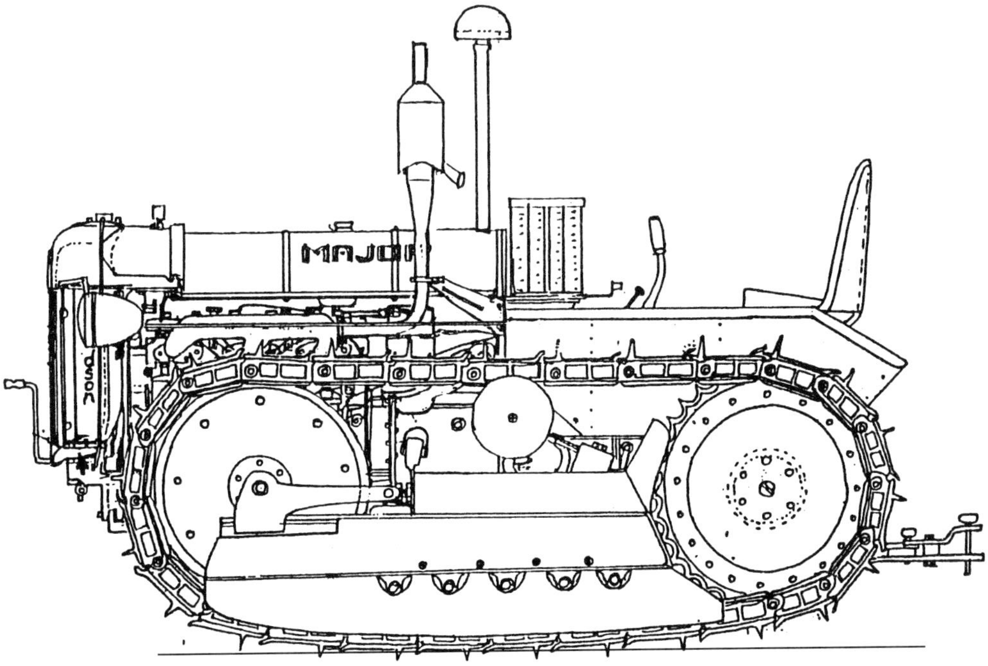

The Competition.

When the Fordson Major first saw the light of day in 1945 Britain was still at War. All tractors supplied to farmers were subject to supply under a War Agricultural Committee permit. Under these arrangements there was 'no competition' as in any case Fordson production accounted for the supply of over 90% of all tractors in the UK from 1939/45. What other manufacturers did was minimal, and the balance was made up by the import of various US makes and models of tractor. Only if the need arose would you get a permit to have anything other than a Fordson.

By VE day on 8th May the E27N was only weeks old; certainly by the time war ended completely on September 2nd it was beginning to get into its stride.

Enter the politicians. The 1945 Labour government policy was one of export! export! export! to earn much needed dollars for repayment of wartime loans. Ford struggled on with a clapped out production line and only half the tractor they had wanted to produce after the War, due to the fact that they could not get material permits to import the machine tools necessary to build a new engine.

Whilst Ford lost out, the need for 'full employment' meant that others were moved into direct competition with the very outfit that had served Britain so well in the war years.

But there were those who were there already. David Brown's production capacity had never been a major threat to Ford, and their production of agricultural machines was very restricted during wartime. Once the peace had been won, they set out improving their models with the eventual introduction of the VAK1C "Cropmaster" in 1947. Whilst production never reached more than 7000 units a year, every sale lost by Ford was one less E27N. The Cropmaster had many more modern features, including six speed gearbox, up-to-date overhead valve engine, and coil ignition.

Marshall of Gainsborough's contribution to the prewar tractor statistics was minimal, but they moved into a postwar world with big ideas. The new Field Marshall had its sights set, in the eyes of the Marshall sales boys at least, on taking on the big fish. Improved the Field Marshall Series One and Two models may have been over the old model 'M', but sales remained firmly in the old camp of contractors and threshermen.

There were of course, still a trickle of imports from the USA; John Deere, Case, Allis Chalmers and Oliver provided a few units in the late forties until import tariffs forced them to either abandon hope or seek a production facility in the United Kingdom. John Deere tried and failed, being again involved in politics where a change of government in 1951 to the Conservatives saw the cutting off of subsidies to those who wanted to set up in the UK. They opted for Germany instead, but it was to be the end of the decade before John Deere would become a threat again.

Allis Chalmers, International, and Massey Harris had a stronger presence in the UK than other American firms. All set up production facilities in Britain in the late nineteen forties, with Government aid. Allis Chalmers opted to build a version of their 'B' tractor at Totton near Southampton, and this uni was never a real challenge to Ford being in the lowe horsepower class and of a design not really suited t the needs of most British farmers. Now if A C had built their 'U' in Britain that would have been another story. The last examples of this august mode reached these shores in 1949 and thereafter Allis ha no real chance of attacking Ford sales.

International had always been a strong contende for a slice of the UK market, and provided the firs real challenger to the E27N in the form of thei Farmall M, built at Doncaster from 1949. This tractor had all the features of the E27N but with a vastly superior engine and gearbox. The one weakness of the M and strength of the E27N were their hydraulic systems. Ford went out from the beginning to provide a good lift with three point linkage, albeit without the Ferguson system, whilst International played around with half baked modifications to their rockshaft lift as fitted to US built tractors. Nevertheless, in the ten years from 1949-59, International provided nearly 40,000 units whilst they could build nearly double that of David Brown, it was still a drop in the ocean if you think that in 1948 over 50,000 E27Ns were built.

Massey Harris sold few tractors in the UK; they were mainly known for machinery. However, the Ground Nut scheme saw the launch of their 744; a 44 frame and transmission fitted with a Perkins P6(TA) engine. Needless to say the E27N with the same engine fitted was a far superior machine, and Massey Harris had to be content with those pickings which one always gets from the minority of customers who will not go along with the masses.

The biggest nigger in the woodpile as far as Dagenham was concerned was always Harry Ferguson. Outside interference at the beginning of the war had caused unnecessary tensions between Detroit and Percival Perry, head of Ford UK. Much as it would have been desirable to see the Ford 9N built at Dagenham, by the time the idea was mooted Britain was at War with Germany, and there was no way that the materials and tooling could have been released for the introduction of a new model.

It was just as well Ford UK never became involved; by the time war ended Ferguson had broken with Ford in the USA and was wooing others in the USA and UK with a view to having his tractor built. The biggest kick in the teeth for Ford was the way in which Government money was plundered on providing production facilities for Ferguson at Banner Lane Coventry. The Ford Salesmen soon dubbed this competition "The Grey Menace", and real competition it was with which Ford had to fight right into the mid Nineteen Fifties. One advantage for Ford was that the original Fergie was petrol only, but soon the TED-20 appeared and closed that gap. Not only this, but once the Psychology of the Irishman and his dealer network had taken over, it was difficult to cure a 'total convert' to the Ferguson system!

Perhaps the tractor which was most like the E27N in size was the Nuffield Universal. Available from the outset in Petrol, Diesel, and TVO forms it had been well known since 1946 that this competition would arrive. It is said that one of the reasons for Ford going 'diesel' was the threat from the Nuffield; both of course used bought in Perkins engines.

Nuffield production figures of the period are not available for comparison, but going by later ones, Nuffields only took a small slice of the market.

The one tractor which proved to be a non starter in the competition stakes was the Turner Yeoman of England. Plagued by problems from start to finish it only took the eventual arrival of the New Fordson Major in 1952 to kill this and other models off, such as the Marshall.

WAC permits continued to be required for the purchase of new tractors until 1948/9. With the emphasis on exports, and 50% of production was exported in the late forties, of all makes, shortages were still evident in the UK market place; it was very much a case of which salesman could get you a tractor first as to who the business went to. There were, of course, plenty of ex War Ag. and MOD. Fordsons coming into the market place and this source very often brought a tractor onto a small farm where horses had reigned supreme until then.

And there was the rub. In those postwar years many sales were to customers who were having their first tractor on the farm. The horse population was to decline rapidly in the years after 1950.

Competition there was for Ford to cope with, and it always seemed that the Dagenham boys had their hands tied behind their backs in the immediate postwar period. At the end of the day, however, it was the Fordson Major which, by sheer volume of production, put Britains farms back into shape in those still austere years from 1945-50. Much as it would have been desirable to have a new model earlier, when it did come, the New Fordson Major was just the machine to take the markets by storm - after all nobody could build tractors in the quantity that Ford could, and at the price! And that price of course also included spare parts - where Ford were always ahead of any competition. In due course a "Vintage Tractor Special" will be available on the New Fordson Major and we can continue the story.

The Backup.

We have all by now realised that any manufacturer getting in on the act and selling tractors in the immediate postwar period had a hard act to follow. And that hard act was the Ford dealer network.

Things were a lot different in the late forties. Ford dealers sold everything.... Cars, trucks, and tractors. Many agricultural merchants were agents for Fordson tractors, almost the same status as a 'retail dealer' today. But the backup which Ford themselves expected from the dealerships had to be total.

Sales, Service and Parts departments all had to operate within guidelines set out by Ford. Not only were staff trained to national standards, but details such as repair charges were set out and standardised throughout the dealer network. Service staff were all expected to attend the courses laid on at Boreham, where they may have benefited from the output of the training given there; the reverse situation occurred when the Ford boys found out the strengths and weaknesses of any particular product from those who had to maintain and repair them.

Needless to say the other manufacturers aped the Ford set-up to a greater or lesser degree. But the sheer scale on which Ford operated meant that there was no way they could really give the same service and at the same prices!

Against a background of the immediate postwar era, when there was an abundance of skilled labour eager to move into 'civvy street' from the forces (where some of them even worked on Ford vehicles of all types), the highly organised, almost military, efficiency of the Ford dealer network made sure that the Ford Anglia, E93A Van, Fordson Major, and the Fordson Thames truck were market leaders. Nothing was impossible for the Ford empire, even to the extent of replacing a complete machine if it failed to please.

Ford were not to be alone for long. Ferguson, with his associations with Ford in the USA set up a comparable organisation to support the sales of the 'Grey Menace'. Those other tractor builders of the immediate postwar period such as International and Massey Harris had good, established dealers able to give almost as good support as the Ford ones. As always, Ford dealers seized any opportunity and offered, as now, repair and servicing facilities for any make of tractor **at Ford prices**!

The E27N was well cared for when it came to Service and repairs. But the age old problem of the farmer, who very often was no mechanic, neglecting that oil change or top-up, or even the use of the grease gun, often found the dealer at a disadvantage when something went wrong. After all, you fed a horse hay and straw and it kept on working; tractors needed a deal more in the way of care if they were to keep in their prime.

When looked after properly, your 'Major' would do its job, but the inevitable lack of maintenance offered to many machines had its effect on those modifications to which the tractor was subjected over its six year production. Ye olde hotplate vaporiser worked fine if the tractor was looked after, yet this one item was the cause of more trouble, along with magneto problems, than any other things which went wrong with the tractor. The need to get rid of that vaporiser brought the improved type in 1950, which in reality created more problems than it solved.

Time moves on, and the simplicity of the E27N with its outdated engine, over exposed and inaccessible driving position, did not endear itself to those used to the good old Fordson 'N'. It did its job however, in most cases, and the tractor world would be the poorer if its presence in the age of preservation is not there, should history have been somewhat different and we might have found a UK built 9N instead! Had that been the case, the cause of the medium powered tractor of 40-50HP would have been put back a decade, after all, out of the E27N came the E27N with P6(TA) and finally the Super Major! A big Fergie never got off the ground, and when the Banner Lane boys did give us a 50HP tractor, Ford were able to answer with a competitor to the MF35. Having said all that, if Henry Ford and Harry Ferguson had not fallen out where might we be today? Of one thing I am certain, there would be none of my favourite tractor, the E27N, to write about.

The Fergie TE-20 was the major threat to the hitherto virtual Ford monopoly of tractor sales in the United Kingdom. By 1951 double the number of Fergie 20s than E27Ns were being built and sold!

David Brown's production could not match either Ford or Ferguson, but the compactness of their product and its more modern design found popularity with some customers who were previously Ford buyers. Here is their 1947 VAK1C "Cropmaster".

Marshall were in a class of their own, with only limited production and sales to dedicated customers. The nail in their coffin came in 1948 when Ford put the P6(TA) into the E27N. By the time the New Diesel Major had been out for a couple of years, UK sales were so bad that the last 3a models had to be sold off to France at reduced prices! Our illustration however shows a series one mark one tractor dating from 1946.

The Nuffield was similar in size to the E27N, but featured a more modern engine, and better gearbox. Fordsons were always cheaper however, and Nuffield's system of fuel management on the TVO tractor somewhat unreliable.

The Massey Harris 744 competed at the top end of the Major scale, that is to say with the E27N fitted with P6(TA) engine. Design weaknesses in the M-H transmission and clutch meant that in direct comparison, the E27N was by far the better machine when Perkins fitted.

The Farmall M had many better features than the E27N, but its hydraulic lift let it down. There were however, around 9000 units a year built, which pecked away at Ford's sales.

The Turner looked promising when launched, but proved a total disaster in service.

The Nuffield Universal with P4(TA) diesel proved a popular alternative to the E27N.

Tractor prices

Allis Chalmers B (TVO) (10/50)	£310.00
Allis Chalmers B (TVO) (10/51)	£335.00
BMB President (10/50)	£239.50
BMB President (10/51)	£284.50
David Brown Cropmaster Model 6VO (10/50)	£456.75
David Brown Cropmaster Model 6VO (10/51)	£477.00
David Brown Cropmaster Model 6SVO (10/50)	£475.75
David Brown Cropmaster Model 6SVO (10/51)	£565.00
David Brown Cropmaster Diesel (10/50)	£633.75
David Brown Cropmaster Diesel (10/51)	£670.00
International Farmall M VO 3 wheel. (10/50)	£645.00
International Farmall M VO 3 wheel (10/51)	£679.00
International Farmall M VO 4 wheel. (10/50)	£637.00
International Farmall M VO 4 wheel (10/51)	£687.00
International Farmall M VO 4 wheel including PTO, Pulley & Lift, (10/50)	£790.00
Ferguson TE-20 (12/46)	£343.00
Ferguson TEA-20 (10/50)	£335.00
Ferguson TED-20 (10/50)	£335.00
Ferguson TED-20 (10/51)	£370.00
Ferguson TEF-20 (10/51)	£490.00
Field Marshall Series 3 (10/50)	£704.00
Field Marshall Series 3 (10/51)	£795.00
Fordson Major Standard Agricultural (6/48)	£237.50 to £226.00
Fordson Major Standard Agricultural (10/50)	£264.00
Fordson Major Standard Agricultural (10/51)	£299.00
Fordson Major Land Utility (6/48)	£296.00 to £281.00
Fordson Major Land Utility (10/50)	£303.75
Fordson Major Land Utility (10/51)	£370.00
Fordson Major Row Crop (6/48)	£255.00 to £242.00
Fordson Major Row Crop (10/50)	£308.50
Fordson Major Row Crop (10/51)	£374.00
Fordson Major Row Crop P6 (10/50)	£599.00
Fordson Major Row Crop P6 (10/51)	£650.00
Massey Harris 744D Standard (10/50)	£854.50
Massey Harris 744D Standard (10/51)	£854.50
Massey Harris 744D Rowcrop (10/50)	£925.00
Massey Harris 744D Rowcrop (10/51)	£925.00
Minneapolis Moline UDS (Dorman 4 cylinder) (9/49)	£1050.00
Nuffield Universal M4 TVO. (2/49)	£495.00
Nuffield Universal M4 TVO. (10/50)	£490.00
Nuffield Universal M4 TVO. (10/51)	£547.50
Nuffield Universal DM4. (10/50)	£657.50
Nuffield Universal DM4. (10/51)	£727.50
Nuffield Universal M3 TVO (2/49)	£487.50
Nuffield Universal M3 TVO. (10/50)	£482.50
Nuffield Universal M3 TVO (10/51)	£547.50
Nuffield Universal DM3 (10/50)	£657.50
Turner "Yeoman of England" (8/49)	£650.00
Turner "Yeoman of England" (10/50)	£695.00
Turner "Yeoman of England" (10/51)	£798.00